Design Principles
in Resource Recovery
Engineering

Design Principles in Resource Recovery Engineering

by Norman L. Hecht

BUTTERWORTHS
Boston • London
Sydney • Wellington • Durban • Toronto

An Ann Arbor Science Book

Ann Arbor Science is an imprint of Butterworth Publishers

Copyright © 1983 by Butterworth Publishers.
All rights reserved.

No part of this publication may be reproduced, stored in a retrieval system, or transmitted, in any form or by any means, electronic, mechanical, photocopying, recording, or otherwise, without the prior written permission of the publisher.

Library of Congress Catalog Card Number 82-46060
ISBN 0-250-40315-3

10 9 8 7 6 5 4 3 2 1

Published by Butterworth Publishers
10 Tower Office Park
Woburn, MA 01801

Printed in the United States of America

This book is dedicated to
my wife, Judy, in appreciation
for her encouragement,
understanding and support.

ABOUT THE AUTHOR

Norman L. Hecht is a Research Scientist at the University of Dayton. He received his Ph.D. degree from the College of Ceramics at Alfred University, State University of New York in 1972 where he also received his M.S. degree in Ceramics Science in 1968, and his B.S. in Ceramic Engineering in 1960.

Dr. Hecht's primary interests lie in the related areas of solid waste management, energy management and materials development. During the past several years Dr. Hecht has served as Principal Investigator on several Environmental Protection Agency projects for the development of improved refuse derived fuels. In addition, Dr. Hecht has coordinated several resource recovery evaluation studies for local communities and for the State of Ohio. In these studies the state of the art for resource recovery has been thoroughly evaluated and the designs for several resource recovery facilities were developed. Dr. Hecht served as chairman of the Technical Advisory Panel to Montgomery County, Ohio and as a member of the Miami Valley Regional Planning Commission.

At the present time, Dr. Hecht is serving as a Co-Manager for the Department of Energy's EADC project to provide assistance to small and medium sized industries in identifying energy conservation measures. In addition, he was a consultant to the City of Dayton Comprehensive Community Energy Management Program sponsored by the Department of Energy to help the city develop long-term energy management policies to promote energy conservation. Dr. Hecht has written a number of articles and technical reports in the environmental, energy, and materials science fields. Since 1960, he has contributed more than 25 articles to the technical literature, has coauthored a book on materials science and contributed several chapters to texts on solid waste management technology.

CONTENTS

Preface xiii
1. Resource Recovery Technology Overview 1
   1.1 Introduction 1
   1.2 Heat Recovery Incineration 2
   1.3 Refuse-Derived Fuels 10
   1.4 Liquid and Gaseous Fuels 10
   1.5 Material Recovery 14
   1.6 Summary 21

2. Municipal Solid Waste 23
   2.1 Introduction 23
   2.2 Sources of Municipal Solid Waste 23
   2.3 Characterization of Components in
       Municipal Solid Waste 24
   2.4 Quantities and Generation Rates for
       Municipal Solid Waste 26
   2.5 Chemical Characteristics of
       Municipal Solid Waste 27
   2.6 Physical Characteristics of
       Municipal Solid Waste 28
   2.7 Combustion Characteristics of
       Municipal Solid Waste 31
   2.8 Industrial and Agricultural Wastes 33
   2.9 Municipal Sludges 33
   2.10 Summary 38

3. Basic Processing Technologies 39
   3.1 Introduction 39
   3.2 Material Size Reduction 39
       3.2.1 Hammermills 40

	3.3	Separation	43
		3.3.1 Size Separation	44
		3.3.2 Density Separation	51
		3.3.3 Magnetic Separation	59
		3.3.4 Other Separation Processes	62
	3.4	Material Handling	63
		3.4.1 Material Transport	63
		3.4.2 Storage	67
		3.4.3 Densification	69
	3.5	Thermal Processes	70
		3.5.1 Combustion	70
		3.5.2 Pyrolysis	71
		3.5.3 Refuse Drying Processes	73
	3.6	Chemical and Biological Processes	74
		3.6.1 Acid Hydrolysis	74
		3.6.2 Cellulose Embrittlement and Enhanced Carbonization	75
		3.6.3 Hydrogenation	75
		3.6.4 Biological Conversion Processes	75
	3.7	Summary	78
4.	Plant Design		79
	4.1	Introduction	79
	4.2	Plant Design Procedures	80
		4.2.1 Procure Basic Data	80
		4.2.2 Analyze Data	80
		4.2.3 Design Production Process	81
		4.2.4 Plan the Material Flow Pattern	81
		4.2.5 Consider General Material Handling Plan	82
		4.2.6 Calculate Equipment Requirements	82
		4.2.7 Plan Individual Work Areas	83
		4.2.8 Select Specific Material Handling Equipment	86
		4.2.9 Coordinate Groups of Related Operations	86
		4.2.10 Design Activity Relationships	86
		4.2.11 Determine Storage Requirements	88
		4.2.12 Plan Service and Auxiliary Activities	88
		4.2.13 Determine Space Requirement	89
		4.2.14 Allocate Activity Areas to Total Space	89
		4.2.15 Consider Building Types	89
		4.2.16 Construct the Master Layout	90
		4.2.17 Evaluate, Adjust and Check Layout with Appropriate Personnel/Obtain Approvals/	

		Install Layout/Follow Up on Implementation	
		of Layout	91
	4.3	Application of Design Procedures for	
		Resource Recovery Facilities	95
		4.3.1 Collect and Analyze Data	95
		4.3.2 Select Process	95
		4.3.3 The Flow Plan	96
		4.3.4 Identify Material Handling Process	97
		4.3.5 Select Equipment	97
		4.3.6 Develop Work Station and Select	
		Material Handling Equipment	99
		4.3.7 Coordinate Groups of Related	
		Operations	101
		4.3.8 Design Activity Interrelationships and	
		Plan Service and Auxiliary Activities	101
		4.3.9 Determine Storage and Space	
		Requirements and Allocate Activities	
		to Total Space	102
		4.3.10 Construct the Master Layout/	
		Establish Building Requirements	103
		4.3.11 Implement Tasks—Reevaluate and	
		Approve Layout, Install Layout	
		and Follow Up on Installation	103
	4.4	Summary	104
5.	Plant Design Exercise	105
	5.1	Introduction	105
	5.2	Community Description and Requirements	106
	5.3	Flow Plan	106
	5.4	Material Balance	112
	5.5	Plant Design	113
		5.5.1 Equipment Selection	113
		5.5.2 Plant Layout	113
		5.5.3 Refuse Receiving Plan	124
	5.6	Site Plan Development	125
	5.7	Update	129
	5.8	Summary	135
6.	Economic Analysis	137
	6.1	Introduction	137
	6.2	Economic Analysis of Proposed Resource	
		Recovery Facility for AMAX, Inc.	137

		6.2.1	Estimated Turnkey Capital Cost	138
		6.2.2	Operating Costs	138
	6.3	Summary		145

Epilogue 149
References 151
Index 157

PREFACE

Resource recovery is an emerging technology that can provide communities with an effective means for solid waste management. The technology incorporates a wide diversity of systems designed to recover useful products from waste. Of primary interest are those systems designed to process the solid waste from the municipal sector of the community. These systems are capable of processing the solid waste from the residential, commercial and institutional sectors, as well as handling the nontoxic (nonhazardous) waste from industry. Systems have been developed in the United States, Canada, Japan and Europe to recover energy, metal, glass, plastics and paper from municipal waste and to convert the organic fraction to humus, protein, alcohol and other chemical products.

Diverse unit processes have been used in these recovery systems. Processes used in mining and mineral recovery, the pulp and paper industry, the coal industry and the food industry have been adopted for various waste recovery functions. In addition, processes have been designed specifically for resource recovery systems. Designs of resource recovery facilities can utilize technologies from all of these disciplines. This book describes the design concepts for a variety of resource recovery systems for processing municipal solid waste.

The objective of this book is to provide the reader with a comprehensive understanding of the major aspects required in the design of a resource recovery system. To accomplish this objective the book has been divided into two parts; Part 1 (Chapters 1 through 3) deals with the technology of resource recovery, and Part 2 (Chapters 4 through 6) deals with the specifics of resource recovery plant design. Chapter 1 presents an overview of the current status of resource recovery, and describes the various commerical systems developed for energy and materials recovery from municipal solid waste. Chapter 2 characterizes municipal solid waste, the raw material for resource recovery. In this chapter, the types composition and properties of municipal solid waste are described.

Chapter 3 presents a detailed description of the variety of unit operations in resource recovery processes. Unit operations for material comminution and separation by density, size, electromagnetic properties, as well as material transport and storage, are discussed.

Having established a basic understanding of the nature of municipal solid waste, the operations for processing the waste and the variety of systems for resource recovery, the second part of this book delves into the basic concepts for designing resource recovery facilities. Chapter 4 deals with the basic concepts of resource recovery plant design. It describes the procedures for developing flow plans and material balances necessary for equipment selection and plant layout planning. In addition, this chapter describes plant support requirements important in the design of resource recovery facilities. Using the information in Chapter 4, a design for a resource recovery facility is developed in Chapter 5. For this plant design exercise, a facility to produce RDF-fluff was selected.

Chapter 6 presents the procedures used for conducting an economic analysis of a resource recovery facility. The procedures for determining capital and operating costs are demonstrated.

The author wishes to acknowledge the efforts of the many individuals who have contributed to the successful completion of this manuscript. In particular I wish to thank my editor, Dr. P. Aarne Vesilind, for his many constructive comments and his patience. I would also like to acknowledge the assistance of Julie Zimmerman for typing the many drafts of the manuscript and Judith Hecht for preparing the index. In addition, I wish to express my appreciation to the support and assistance provided by my employer, the University of Dayton.

Norman L. Hecht

CHAPTER 1

RESOURCE RECOVERY TECHNOLOGY OVERVIEW

1.1 INTRODUCTION

Conventional practice has been either to burn or to bury collected solid wastes. However, because of the many problems associated with landfill and incineration, plus the greater concern for improving environmental quality and the need for conserving our energy and mineral resources, a number of new techniques for improved solid waste management are being developed. Most are concerned with recovering and utilizing the valuable materials in solid waste.

Resource recovery is the term applied to the processes and systems designed to recover components in waste and convert them to useful products. Systems have been developed for energy recovery; compost production; wood fiber, glass and metal recovery; and alcohol and protein production. However, major emphasis has been on those processes designed to recover the thermal energy in the solid waste. To date the most effective means for recovering this thermal energy has been the utilization of waste as a fuel. Solid waste can be processed into a number of different fuel forms and used in a variety of furnaces, e.g., it can be used in the raw or "as-received" condition or processed into a solid, liquid or gaseous fuel.

Laboratory, pilot-scale and full-scale facilities have been developed for a wide range of resource recovery processes. This chapter presents a brief review of the major resource recovery technologies commercially available.

1.2 HEAT RECOVERY INCINERATION

In the United States, energy recovery has been the primary goal for most of the resource recovery processes developed to date. A majority of these processes are designed to utilize refuse as a solid fuel, and one of the most common approaches is to mass burn or incinerate the as-received refuse to generate steam or hot water. The steam produced can be used for space heating and cooling (district heating systems), industrial processes and electric generation. In most of these systems very little processing is employed either before or after combustion.

Although heat recovery incinerators have a long history in both Europe and the United States, they were not used on a large scale for processing municipal waste until after World War II. In Europe, large-scale steam-generating incinerators have been in operation for more than 20 years. About 20-30% of European refuse is processed in heat recovery incinerators (over 300 units). In Japan, large-scale heat recovery incinerators have been used for the past ten years. In the United States, the use of heat recovery incinerators is not widespread, although in the past few years several municipal incinerators with heat recovery systems have been built. In a 1981 report, 18 heat recovery incinerator units were reported operational and 16 were in various stages of construction [1].

Mass burning units combine waste combustion with heat recovery to produce steam or hot water. The basic components of the incinerator and heat recovery unit are shown schematically in Figure 1-1. Incinerators can operate on a continuous and/or periodic batch basis. Continuous-feed incinerators, e.g., the traveling-grate, reciprocating-grate, ram-feed and rotary-kiln, are commonly used for municipal incineration. The heat of combustion can be recovered for steam generation by using a water wall system in place of refractory walls in the combustion chamber, or a boiler unit located after the combustion chamber. A typical water wall incinerator system commonly used in the United States and Europe for steam generation is shown schematically in Figure 1-2 [2].

In most systems, refuse is received in a storage pit and transferred to the incinerator by an overhead crane and clamshell grab. The storage pit or bunker normally provides several days' storage capacity for the facility. The grab deposits the refuse into the furnace feed or charging chute and then the refuse is normally gravity fed into the furnace. The charging hoppers are usually water cooled and are always kept filled to capacity to prevent hot gases from entering the refuse storage area. Most incinerators are operated under negative pressure to prevent hot gases and odors from escaping from the system.

A variety of furnace designs have been developed for burning refuse.

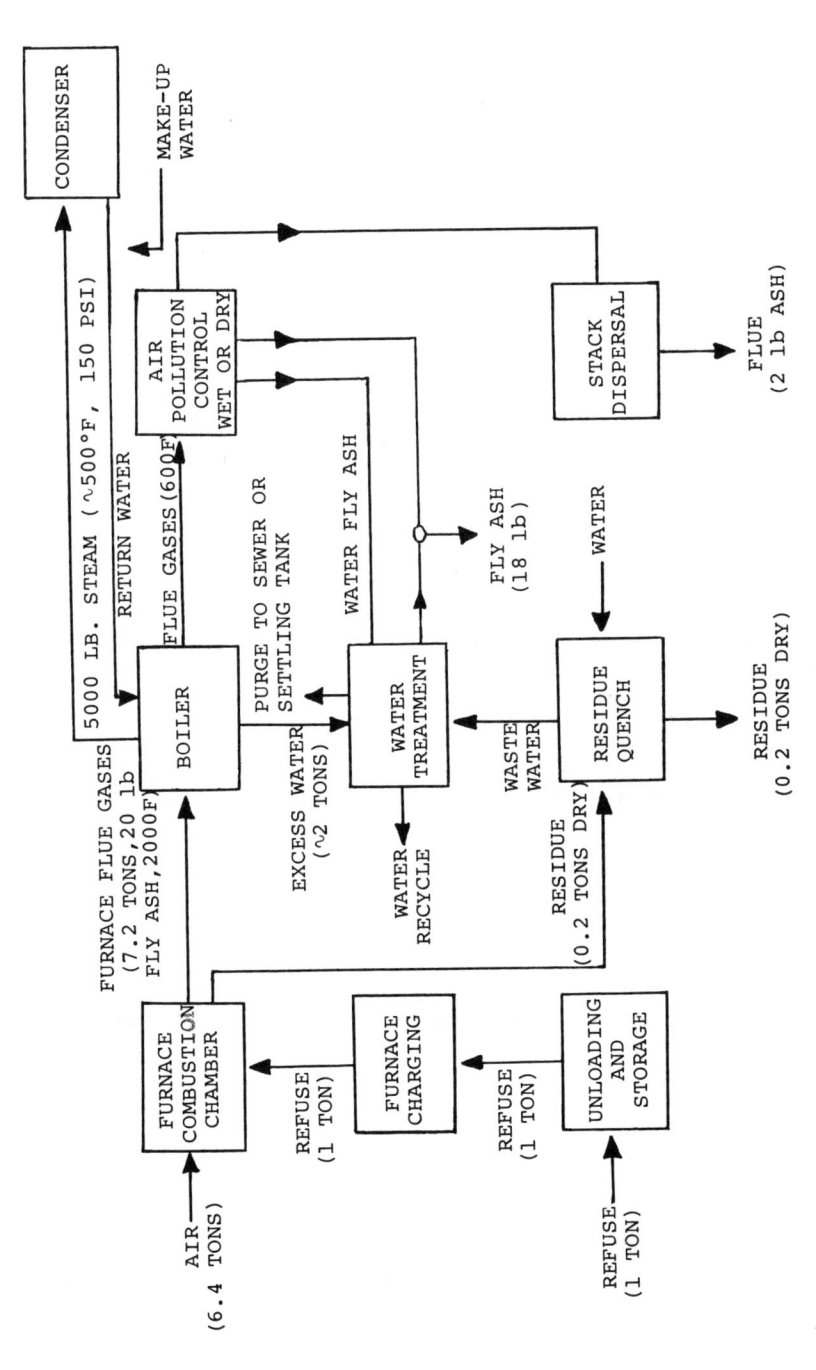

Figure 1-1. Basic components of a heat recovery incinerator. *NOTE*: quantities in parentheses are rough measures of flowrates and temperatures.

4 DESIGN PRINCIPLES

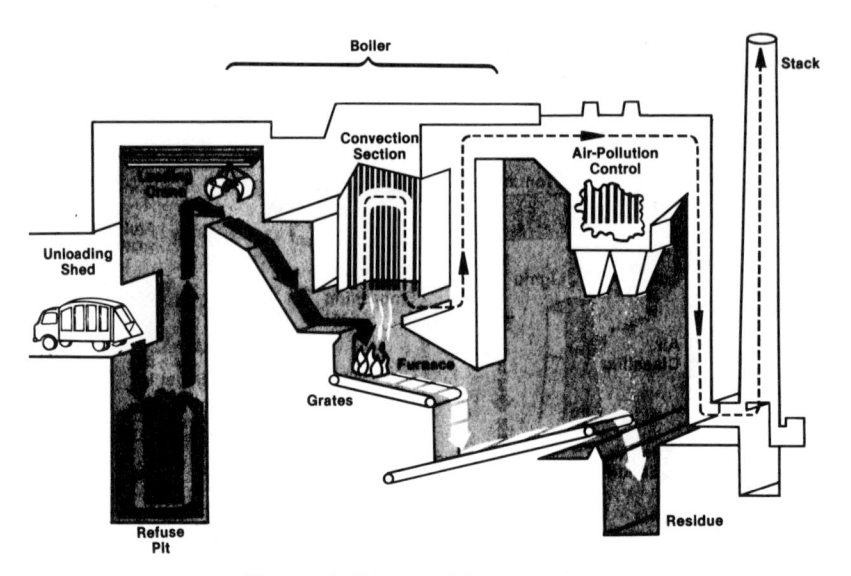

Figure 1-2. Water wall incinerator [2].

Most of the units are designed for mass burning of the raw refuse, and no special processing is required. The refuse is conveyed through the furnace by some type of stoker system and then dried and burned on the stoker unit, with air being introduced from both under the stoker and over the refuse. In water wall units a minimum of excess air is required. The residue from combustion normally is carried by the stoker to a water quencher, and the quench pit can serve as a water seal to the furnace.

The stoker unit is required not only to convey the refuse through the furnace, but also to agitate the refuse, permitting more complete combustion. Several furnace designs include auxiliary burners to assist in combustion of the refuse. This may be required in cases in which wet refuse is frequently encountered.

The residue from refuse incineration is usually a wet complex mixture of metal, glass, slag, charred and unburned organics, and ash. However, in some incinerator units the residue is collected dry. This residue (wet or dry) usually is disposed of in a landfill. In some units the residue is passed through a rotary screen, which separates the metal from the ash; the metal is recovered and the ash is used as a roadbed fill. The United States Bureau of Mines (BOM) developed a process that recovers metal and glass from incinerator residue [3]; however, this system has not been commercially proven.

The hot gases leaving the furnace chamber contain the products of combustion, particulates and other gases released by thermal decomposition of the raw refuse. These hot gases pass through the boiler section and are then treated in a gas cleaning unit. Air pollution codes require the removal of particulates from the gas before they can be released to the atmosphere. Although a number of different types of particulate removal systems are available, electrostatic precipitators are reported to be the most effective. Reports to date indicate that these precipitator units achieve 95-99% efficiency and meet the required U.S. Environmental Protection Agency (EPA) performance levels. The fly ash collected is usually added to the incinerator residue for disposal.

In an electrostatic precipitator, solid particles in the flue gas are electrostatically charged by a high-voltage discharge, and the gas is passed through a high-voltage electrostatic field where the solid particles are attracted to a positively charged collection surface. The collection particles are either shaken or hosed down from the collecting surface periodically. Optimum operating temperatures for the precipitators are between 450°F and 550°F. Flue gases from incinerators without boiler units can average from 1200° to 1400°F, and it is necessary to cool these gases prior to gas cleaning.

1.3 REFUSE-DERIVED FUELS

A number of mechanical, thermal, chemical and biological processes have been developed for recovering the combustible fraction in municipal solid waste. In most of these processes, the organic fraction is converted to a processed solid fuel compatible with conventional coal-fired combustion systems. The fuel obtained is usually termed "refuse-derived fuel" (RDF) and its specific characteristics depend on the particular process employed.

The most common sequence of processing used for RDF production is the one developed in an EPA demonstration program with the city of St. Louis and the Union Electric Company [4]. The system consists of a series of processes that result in the production of a fuel termed RDF-"fluff." The process consists of primary shredding for homogenization and size reduction (minus 6 in.), air classification of the shredded refuse to separate the combustible light fraction from the predominantly noncombustible heavies, magnetic separation of the ferrous from the heavies fraction, screening to remove dirt, glass and grit from the light fraction and secondary shredding to reduce the fuel fraction to minus 2 in. This processing sequence sometimes also is termed "front end" resource recovery. A flow plan of the St. Louis demonstration is shown in Figure 1-3. The primary product from this type of processing procedure is a fluff

6 DESIGN PRINCIPLES

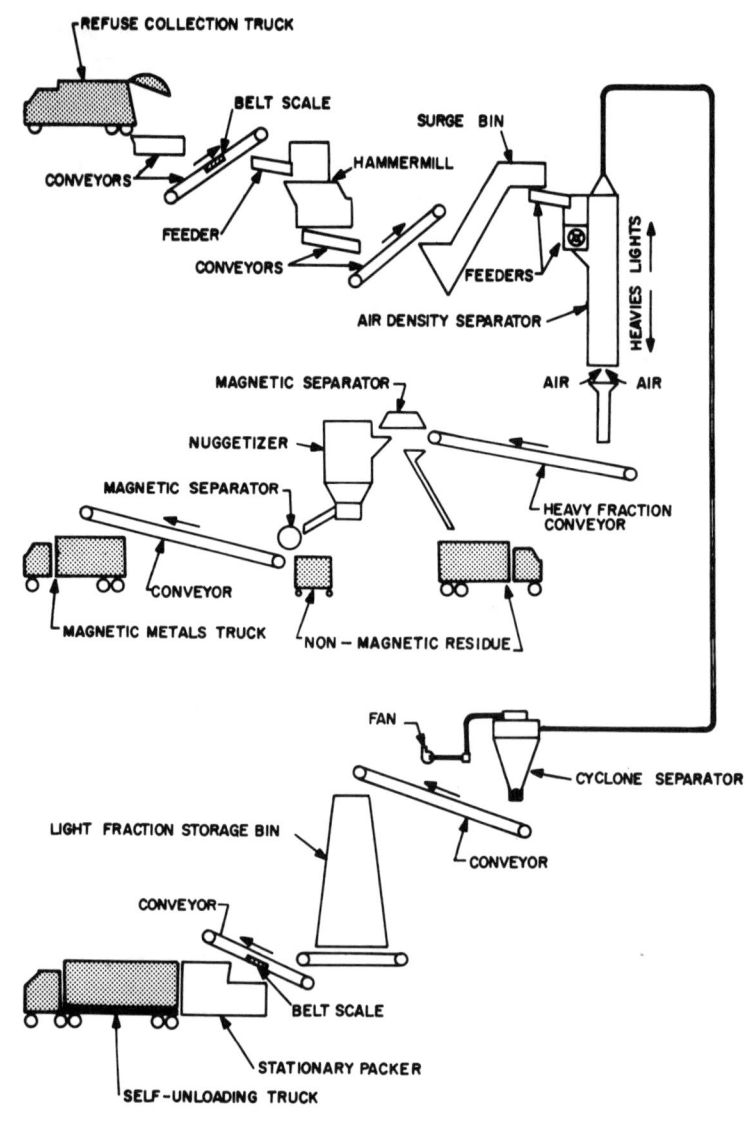

Figure 1-3. Processing plant for RDF-fluff based on the St. Louis project [4].

fuel (minus 2 in.) material similar to that collected in a vacuum cleaner and in ferrous metal. In some plants, additional processing equipment is included to permit recovery of glass and nonferrous material from the heavies. Although each plant may have a somewhat different processing procedure, the fuel fraction is primarily the shredded organic materials, with a small percentage of noncombustible material (glass, bones, ash, metal, etc.).

Since the development of this design more data from operating plants have been obtained, and it is felt that the original design should be modified to include a trommel as the first processing step (similar to the recource recovery plant in New Orleans) [5] to achieve more effective recovery. The basis for this decision is discussed more fully in Chapter 4.

The RDF-fluff can be fired in both industrial and utility-class boilers and in selected industrial furnaces. The fluff normally is used as a partial replacement of coal (10-30% heating value basis) but can be used as the only fuel for a boiler unit (dedicated facility). To date the best performances are reported for the fluff used in moving grate or spreader stoker furnace units (semisuspension). The RDF prepared by dry mechanical processing has about 5000 Btu/lb, 20% moisture and 20% ash.

Full-scale plants in Ames, Iowa, Milwaukee, Wisconsin and Chicago, Illinois have been in operation for the past several years. As of early 1981, there were some 18 plants either in operation or under construction [1]; however, three of these facilities (Milwaukee, Chicago and Lane Co., Oregon) have been closed temporarily due to operating problems. In addition, a number of communities are in various stages of planning resource recovery systems for the production of RDF-fluff. Although the technology for RDF-fluff production is emerging around the United States and in several European countries, it is important to recognize that a number of problems need to be resolved and technical developments made to permit more widespread implementation and greater economic effectiveness.

In addition to dry mechanical processes for recovering RDF, the Black Clawson Company has developed a wet process based on paper-making technology, which is used to recover a fuel fraction [6]. This hydrapulping process was initially designed for recovery of paper fibers. However, it can be employed to separate and recover more than 90% of the organic fraction in municipal solid waste (MSW). A flow plan for this process is shown in Figure 1-4 [7]. As shown, refuse is taken from the receiving area and introduced into the hydrapulper, where it is suspended in water and mechanically broken down in size. The slurry from the hydrapulper is mechanically dewatered to about 50% solids content. This sludge product has been fired in wood waste-type boilers. It also can be pelletized or powdered for use in suspension-fired boilers.

The Black Clawson process was demonstrated in a 150-ton/day facility for the city of Franklin, Ohio. The Black Clawson process also is being used in a Hempstead, New York resource recovery plant that processes 2000 ton/day. The fuel produced is used in a dedicated boiler to generate

8 DESIGN PRINCIPLES

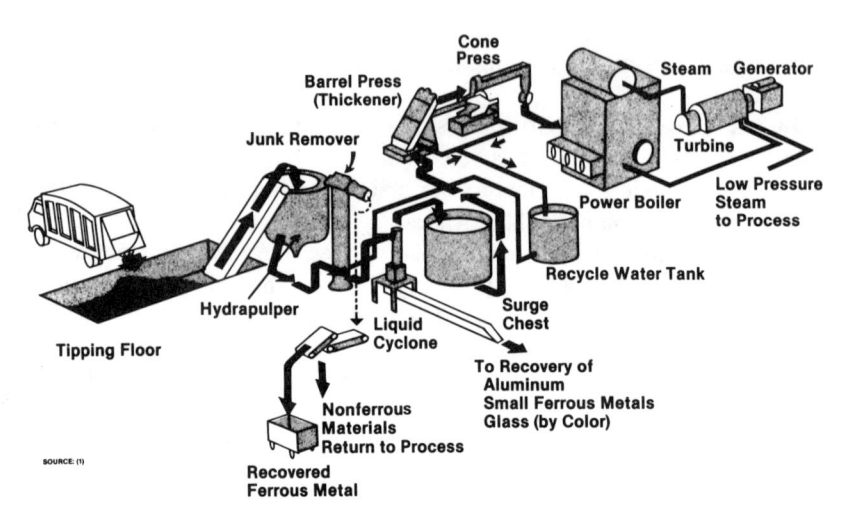

Figure 1-4. Flow plan for a wet resource recovery process [7].

steam for electric production. A similar plant to process 3000 ton/day is also being built in Dade County, Florida. (The operational status of both the Hempstead and Dade County facilities was in question as of late 1982.)

Densified refuse has been used as a fuel for a number of years [8,9] with varied success. Cubes, pellets, briquettes, buttons, etc., of refuse have been tried in a variety of demonstration programs. The demonstration program at Wright-Patterson Air Force Base probably has been the most extensive investigation for the use of densified refuse-derived fuel (D-RDF).

Most of the studies reported in the literature have been with spreader stoker (semisuspension) boiler units. D-RDF has been used to replace from 10% to 100% coal on a heat content basis. Based on the information to date, it appears that the D-RFD has proven to be a satisfactory fuel for stoker-fired boiler units. The D-RDF is usually prepared by densifying finely shredded refuse from which metal, glass and other contaminants have been removed. The refuse is usually densified by compaction (cubes, etc.) or extrusion (pellets, etc.) equipment. A variety of different processes have been developed for the production of D-RDF and, as would be expected, they display a wide range of property characteristics. The densities of the pellets range from 75 to 85 lb/ft^3 with bulk densities of 25 to 35 lb/ft. Heat content varies from 5000 to 8000 Btu/lb ash content from 10 to 30% and moisture content from 10 to 15% [8].

The schematics for two different D-RDF flow plan concepts to process 100 ton/day are shown in Figure 1-5. Flow plan #1 incorporates primary

RESOURCE RECOVERY TECHNOLOGY OVERVIEW 9

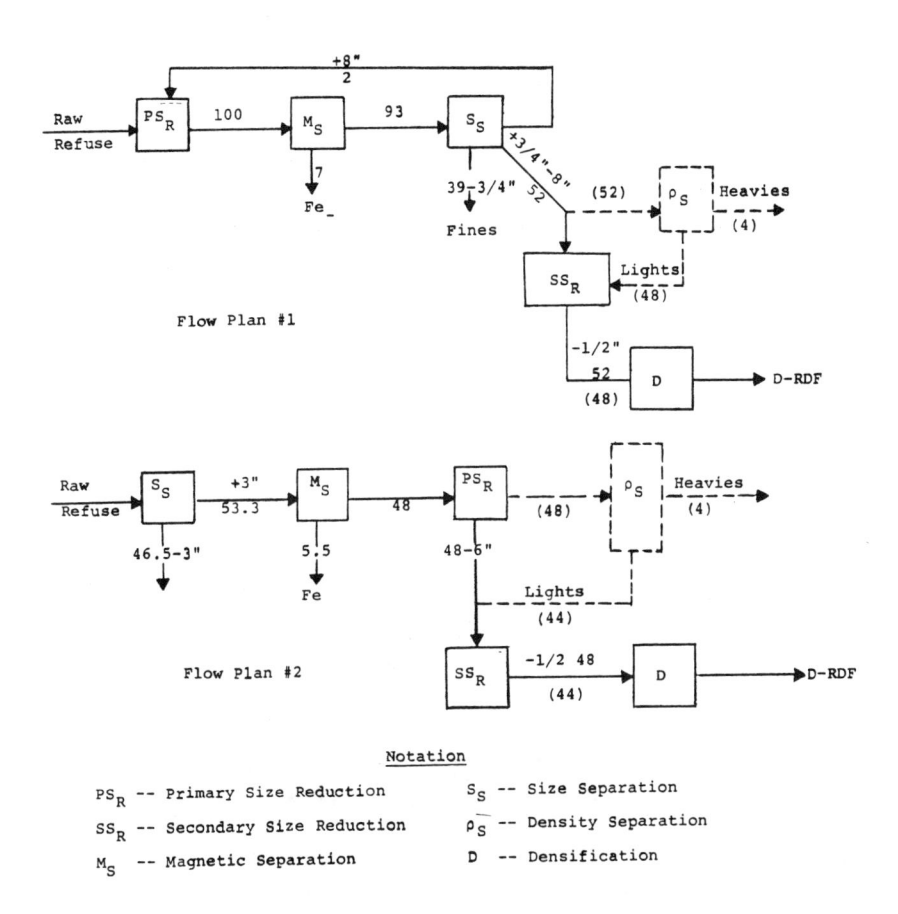

Figure 1-5. Possible D-RDF flow plans.

Notation:

PS$_R$ -- Primary Size Reduction   S$_S$ -- Size Separation
SS$_R$ -- Secondary Size Reduction ρ_S -- Density Separation
M$_S$ -- Magnetic Separation   D -- Densification

size reduction, magnetic separation, size separation, secondary size reduction and densification. In flow plan #2, the initial processing step is size separation followed by magnetic separation, and then primary size reduction of the larger-sized material followed by secondary size reduction and densification of the shredded material. An intermediate step of density separation to remove the heavier noncombustible materials prior to secondary size reduction can be considered also.

In addition to densification, conversion of RDF to a powder to improve fuel quality is being pursued. Converting the organic fraction of refuse into a finely powdered fuel offers a number of advantages. In powdered form the refuse is more compatible with the fuel used in suspension-fired boilers; it is more easily slurred with oil for firing in liquid fuel units; and it is more easily pelletized for use in moving-grate

10 DESIGN PRINCIPLES

units. In addition, the powder can be used as a filler in rubber and plastics or as a feedstock for gasification or liquefaction.

A number of thermal and chemical treatments have been identified that promote the conversion of the organic fraction in refuse to a fine powder. Most of the processes chosen embrittle or degrade the waste.

A proprietary process for recovering a powdered fuel has been developed by Combustion Equipment Associates (CEA). The air-classified, shredded-light fraction is chemically treated with a mineral acid during a hot ball milling process. The ball-milled material is screened to recover the 20 mesh fraction. This fine powder contains about 15% inert material, which can be reduced by a second air classification treatment. The powdered fuel produced by CEA is marketed under the trade name Eco-Fuel II (Figure 1-6). This material has a reported heating value of 7500–8000 Btu/lb, a density of 30 lb/ft^3, a moisture content of approximately 2% and an ash content of approximately 10% [10].

Although a variety of refuse-derived fuels have been produced, the performance of these fuels in combustion units has been quite varied. In some cases the fuel has been very effective; in others, the fuel has not been very satisfactory. Higher-quality (low ash, low moisture, high Btu) and greater consistency (size distribution, composition, etc.) would improve combustion performance and fuel marketability. The major problems associated with the use of RDF have been difficulty in handling (clogging, erosion, etc.), slagging and incomplete combustion, high ash residues and some emission problems (lower precipitator performance).

The quality of the fuel obtained is dependent to a large extent on the effectiveness of the processing system. The ability of the processing system to handle the heterogeneous refuse effectively determines the technical and economic efficiency of the system. To date a number of the mechanical processing systems developed for RDF recovery have been hampered by technical problems. The primary difficulty is the limited experience with these processing systems and the lack of long-term experimental data for the specific unit processes with municipal wastes. With time these problems will be resolved.

1.4 LIQUID AND GASEOUS FUELS

A number of thermal, biological and chemical processes have been developed to produce both liquid and gaseous fuels from refuse. Pyrolysis, liquefaction and bioconversion are the dominant processes for converting cellulose wastes to liquid or gaseous fuels. Hydrolysis by chemical or biological treatment is used for the conversion of cellulose

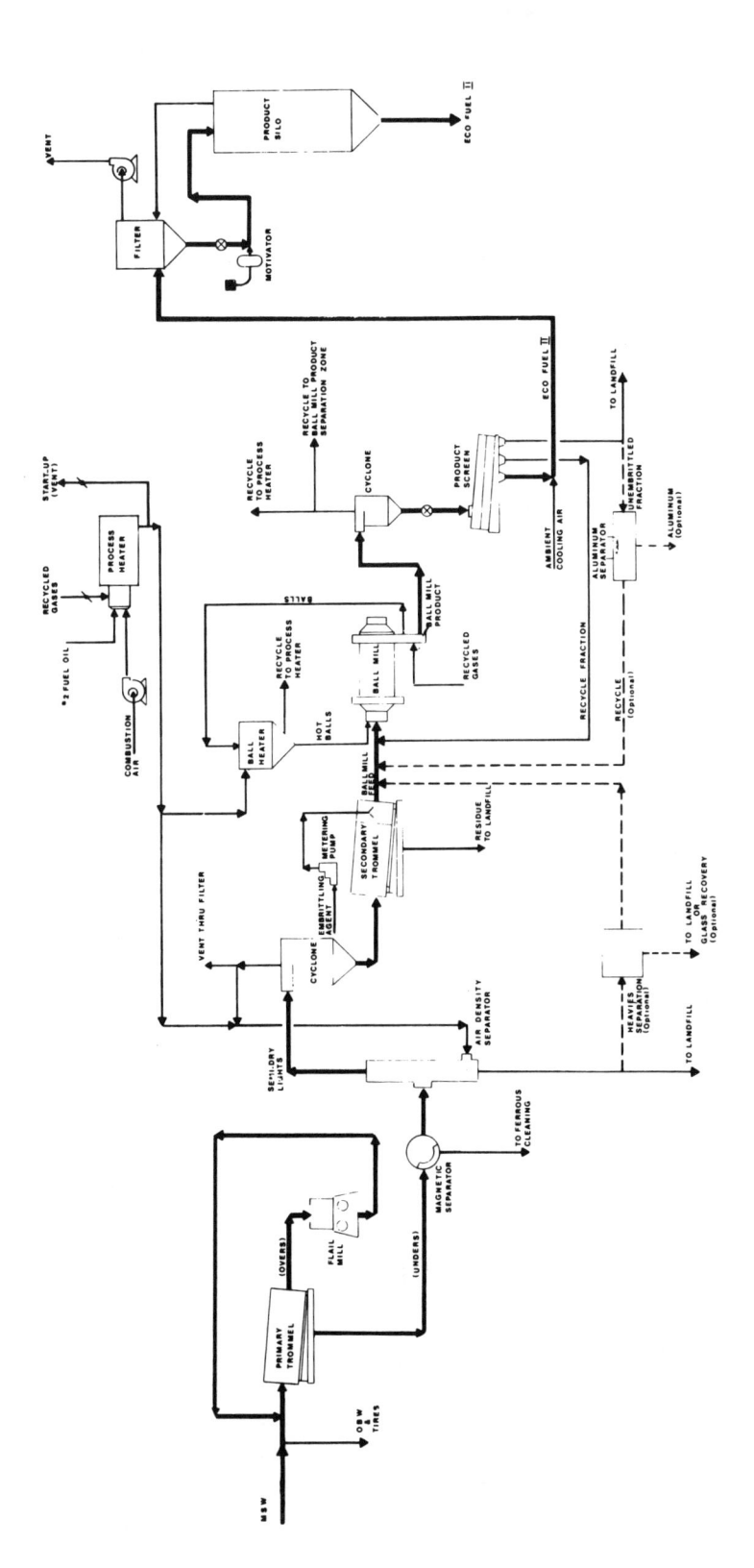

Figure 1-6. Flow plan for the Eco-Fuel II process [10].

wastes to glucose as a preliminary step for the production of alcohol by fermentation. At this time, all these processes are at the laboratory, pilot or demonstration level. No full-scale facilities are commercially available for liquid or gas production from refuse.

One of the advanced types of technologies for fuel production is pyrolysis. Thermal decomposition of refuse in the absence or partial absence of oxygen is used in a variety of pyrolysis processes to produce liquids, gases and char. The quality and quantity of the bitumen-like liquid, gas and char generated will vary and is a function of the time-temperature sequence used for each particular process.

Several pyrolysis processes have reached the demonstration stage of development. Most of the processes developed have encounted a number of difficulties and have been discontinued. A typical pyrolysis process is the one developed by the Occidental Research Corporation and was evaluated in a demonstration program in El Cajon, California (San Diego County) (Figure 1-7). The system uses a flash pyrolysis process that converts finely shredded, air-classified refuse into a combustible liquid, a medium-Btu fuel gas and a char residue. The process is designed to maximize the liquid fuel yield and the char, and fuel gas is recycled in the system. This system also encountered serious processing problems and was closed in 1979 [12].

The Torrax®, process, developed by the Carborundum Company, is a high-temeprature, air-fed slagging pyrolysis process that converts refuse to a low heating value fuel gas, char and a fused frit. High process temperatures permit fusion of the inert fraction of the refuse. Superheated air introduced into the furnace supplies the thermal energy for slagging of the noncombustibles. Natural gas or a fraction of the pyrolysis gas can be used to preheat the air. The low heating value of the pyrolysis gas (150 Btu/scf) requires that it be used for steam generation in an adjacent boiler unit. The carbon char recovered during the pyrolysis gas cleaning stage can be reinjected into the gas at the boiler as a means of increasing the heating value of the gas and improving the energy recovery efficiency of the Torrax process. This process is being evaluated in several European facilities [13].

In addition, a number of other processes are being developed and studied. Rotary kilns, vertical and horizontal shaft furnaces, fluidized bed furnaces and batch-type reactors are being employed for fuel production. Within the next several years a number of these systems may become commercially available.

Liquefaction is a means for obtaining liquid fuels from cellulose by the combination of chemical and thermal treatments. A number of these treatments have been used for processing the organic fraction of MSW.

RESOURCE RECOVERY TECHNOLOGY OVERVIEW 13

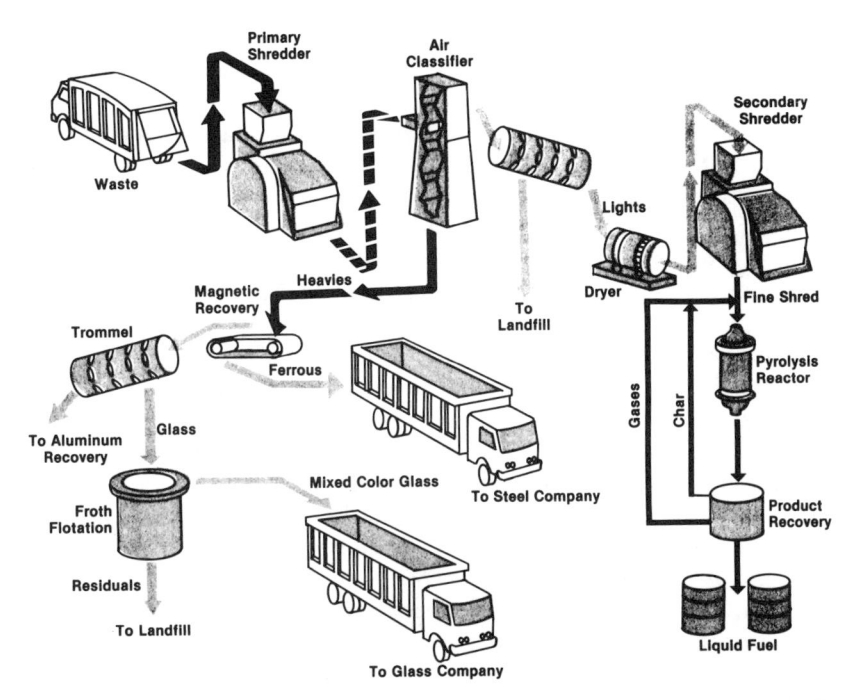

Figure 1-7. Production of liquid fuel from solid waste using the Occidental process [11].

The conversion of cellulose to liquid products is not new and dates back to the early 1900s.

Most commonly, cellulose products were hydrogenated at temperatures from 300 to 930°F at elevated pressures and in the presence of a catalyst. Two recently reported processes for the liquefaction of municipal waste were developed by the Bureau of Mines and Worcester Polytechnic Institute (WPI) [14,15].

In the Bureau of Mines process, cellulose waste is treated with carbon monoxide and water to obtain a liquid fuel. The reaction is carried out in an autoclave at temperatures of 480–750°F and pressures of 1400–4200 psia. Sodium carbonate is frequently used as a catalyst for this reaction; however, the alkali salts present in municipal refuse also can serve as the catalyst. A small-scale demonstration project is underway to evaluate the feasibility of this process in Albany, Oregon [14].

The Worcester Polytechnic Institute has devleoped a laboratory process that converts municipal solid waste to a usable liquid fuel. In this process, shredded refuse is slurried in paraffin oil and mixed with a 0.2% nickel hydroxide catalyst. The slurry is pressurized with hydrogen to 1000 psig in an autoclave and heated to 800°F. This hydrogenation reaction in the

14 DESIGN PRINCIPLES

presence of a nickel catalyst results in the production of oils (bitumens) from the cellulose products in refuse. A schematic of the flow plan for this process is shown in Figure 1-8 [15].

A number of anaerobic digestion processes also have been developed for generating methane gas from solid organic wastes. In many of these processes, municipal refuse and sewage sludge can be combined in the digester. Maximum energy recovery is less than 50% because more than half the organics are not digestible. The residue produced must be disposed of or used in some other type of fuel process.

In the anaerobic digestion process, MSW is usually size reduced (shredding or hydrapulping) and the metal and glass removed. The organic fraction then is mixed with sewage sludge, which contains the microbes for methane generation from refuse. The slurry then is transported to digestion tanks for gas production. Gaseous products are produced continuously and must be bled from the tanks, and the gas obtained must be processed to remove carbon dioxide, hydrogen sulfide and water. The dried gas then can be compressed and introduced into the pipeline. A demonstration of the process is underway at Pompano Beach, Florida, sponsored by the Department of Energy (DOE). A flow plan of this process is presented in Figure 1-9 [16,17].

Methane production by anaerobic digestion is also a process that is naturally occurring in most landfills. In essence, a landfill can be considered to be a large anaerobic digester. The usual practice is to vent the gases produced to prevent explosions. However, the gas generated in a landfill can be collected and utilized. The first commercial landfill methane recovery facility was opened by the Los Angeles County Sanitation District in 1971. By 1979 some 16 landfills had projects for recovering methane. The gas is obtained by drilling wells into selected sites on the landfill using the same techniques used in the extraction of ground water. The gas recovered is upgraded by removing the CO_2 and moisture and compressed for introduction into the pipeline [18].

1.5 MATERIAL RECOVERY

In addition to energy recovery (as steam or fuel), a number of other products can be recovered from municipal refuse. Metal, glass, plastics and paper can be recovered directly from MSW by a variety of mechanical processes. In addition, compost, cellulose fiber and a variety of cellulose-derived chemicals can be obtained by thermal chemical and biological processing of the organic fraction in municipal refuse. Metal and glass usually are recovered as by-products during fuel or celluose

RESOURCE RECOVERY TECHNOLOGY OVERVIEW 15

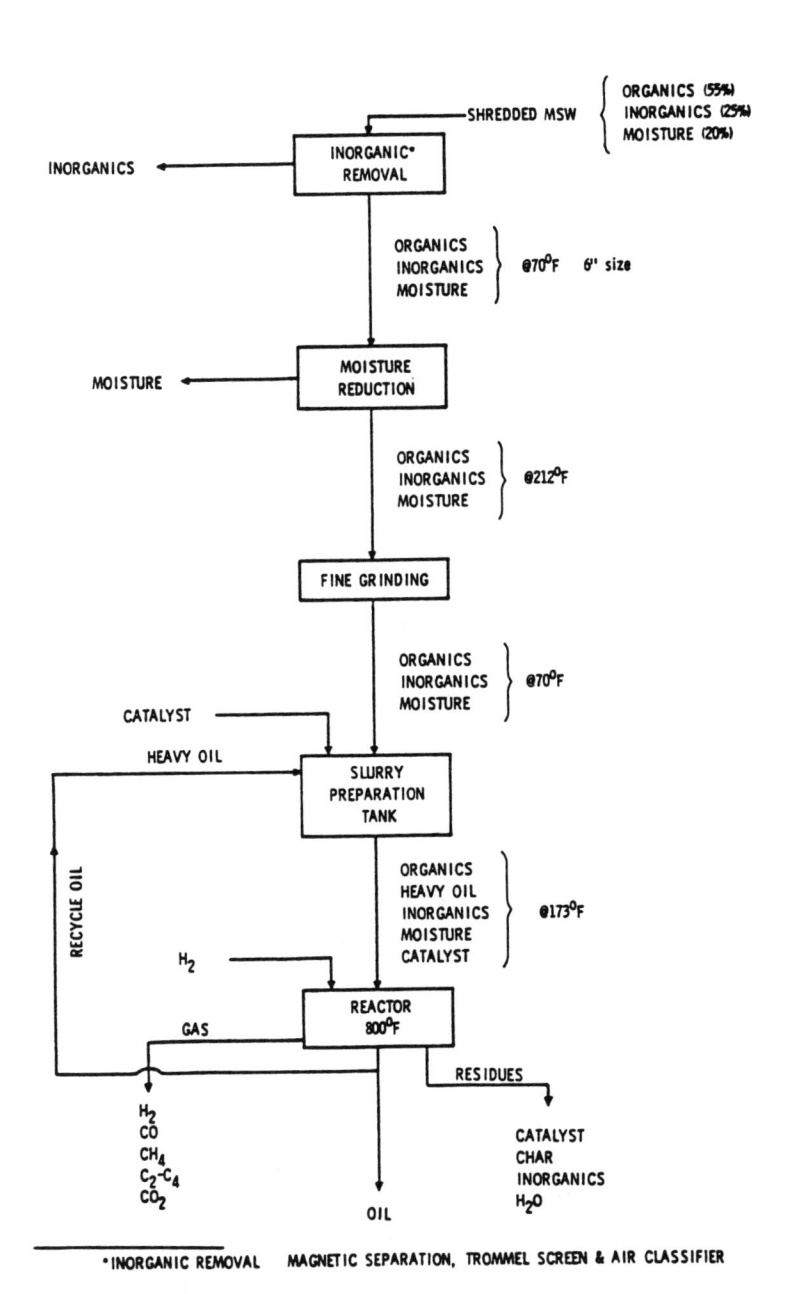

Figure 1-8. Flow plan for the WPI process [15].

16 DESIGN PRINCIPLES

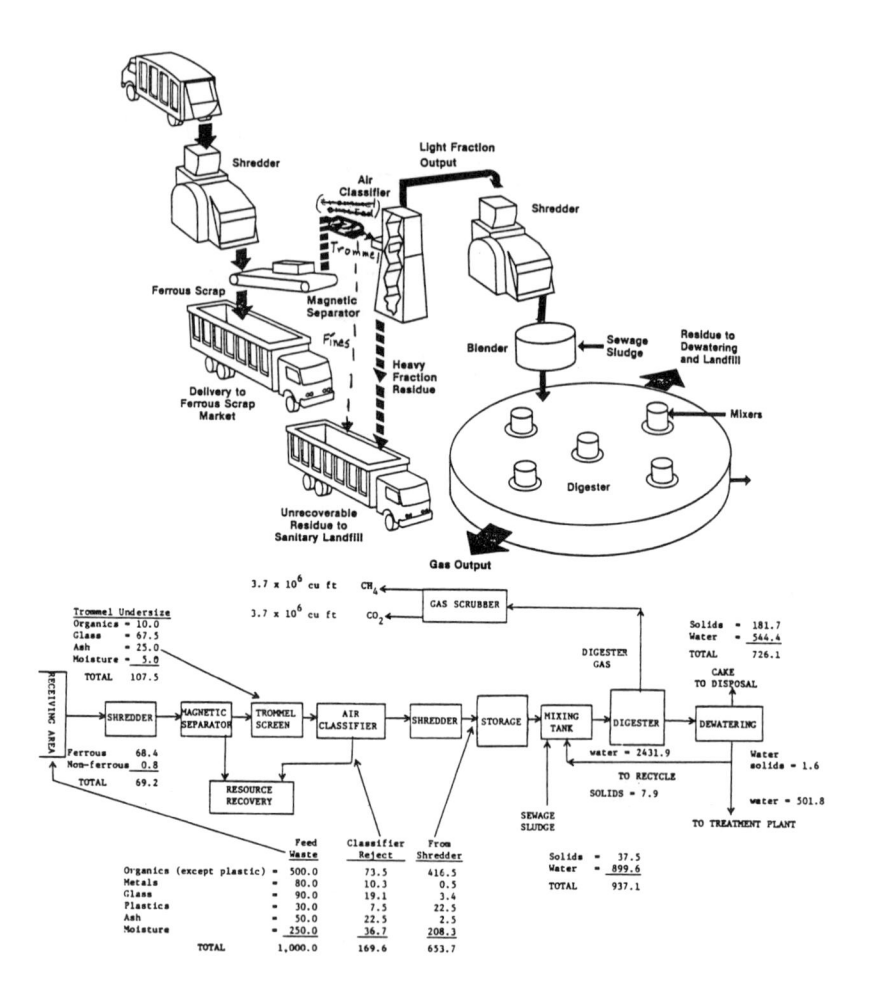

Figure 1-9. Anaerobic digestion plant, Pompano Beach [16,17].

fiber recovery. In these processes, the raw refuse usually is shredded and density separated. A trommel with 5-in. openings can be used prior to shredding to concentrate the glass and metal fractions.

After separation from the light organics, the ferrous metal is recovered by magnetic separation, while a variety of sizing, density and conductivity devices are used to recover the aluminum and glass in the heavy concentrates. An example of this is the Bureau of Mines process shown in Figure 1-10 [19]. The larger glass pieces also can be recovered by optical sorting equipment. A more detailed discussion of the specific unit

RESOURCE RECOVERY TECHNOLOGY OVERVIEW   17

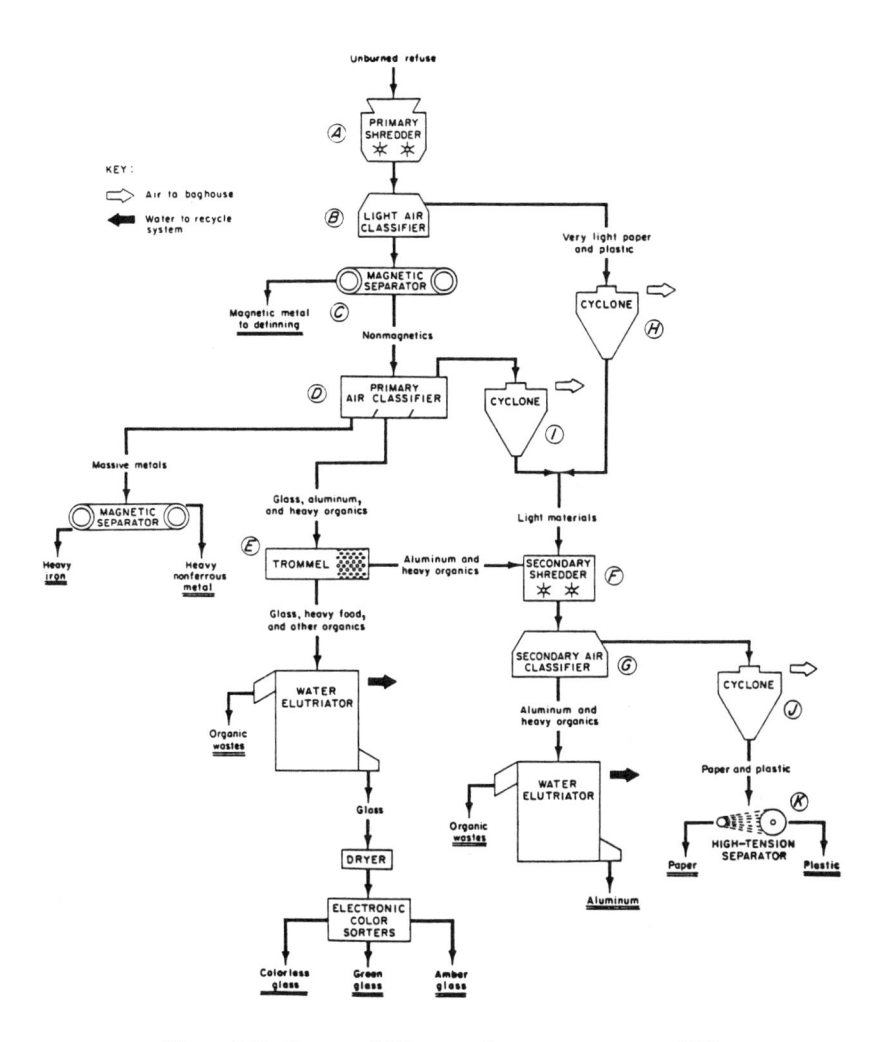

Figure 1-10. Bureau of Mines product recovery system [19].

operations for metal and glass recovery is presented in Chapter 3. The market for glass and ferrous metal recovered from MSW has not been very good and, at present, it does not appear that recovery of these products will be a very dominant part of a resource recovery system.

Several processes also have been developed for paper recovery. Most of the dry processes use air classification and additional processing to obtain a high-quality paper product. Processes of note have been developed by the Flakt Company of Sweden and the United States Forests Product Laboratory. A flow plan of the Flakt process is presented in Figure 1-11

18 DESIGN PRINCIPLES

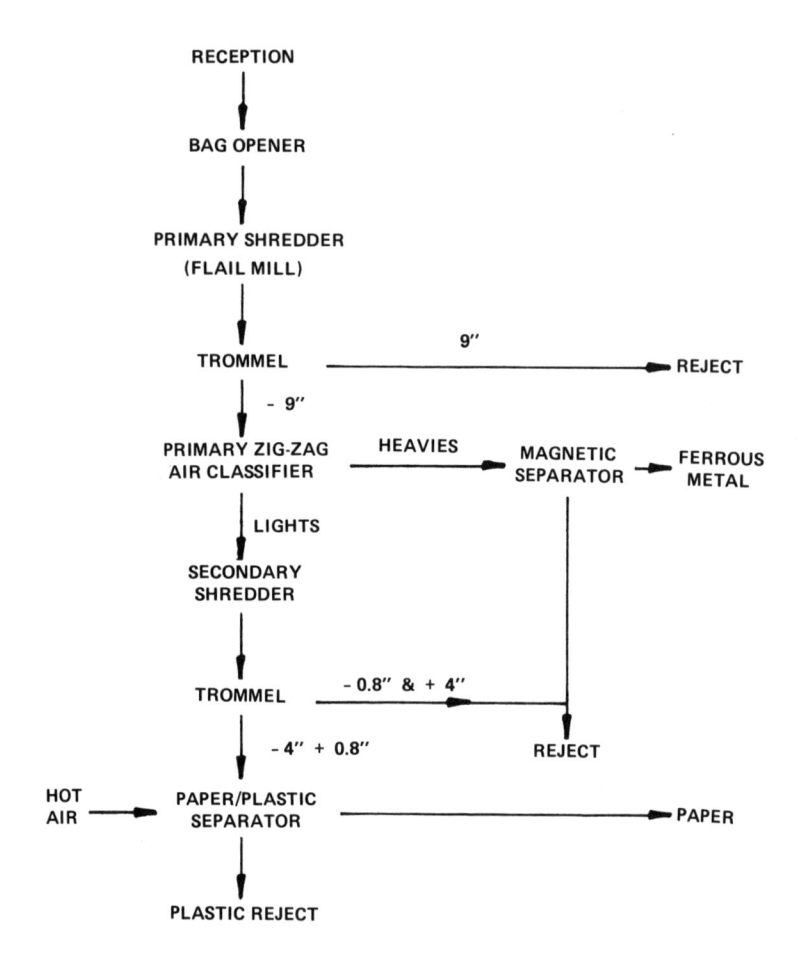

Figure 1-11. Flakt system—simplified flowsheet [20].

[20]. The most common wet process for paper fiber recovery is the one developed by the Black Clawson Company and demonstrated in Franklin, Ohio. This process was described in the discussion on RDF processes, and a flow plan is presented in Figure 1-12. Film plastics usually are recovered as a by-product from the paper recovery processes. In addition, a number of processes have been developed in the United States, Europe and Japan for recovery of mixed plastics. However, secondary products prepared from mixed plastics have not been very effective to date.

Biological and chemical processes have been developed for converting the organic fraction of MSW to humus or commercial chemicals.

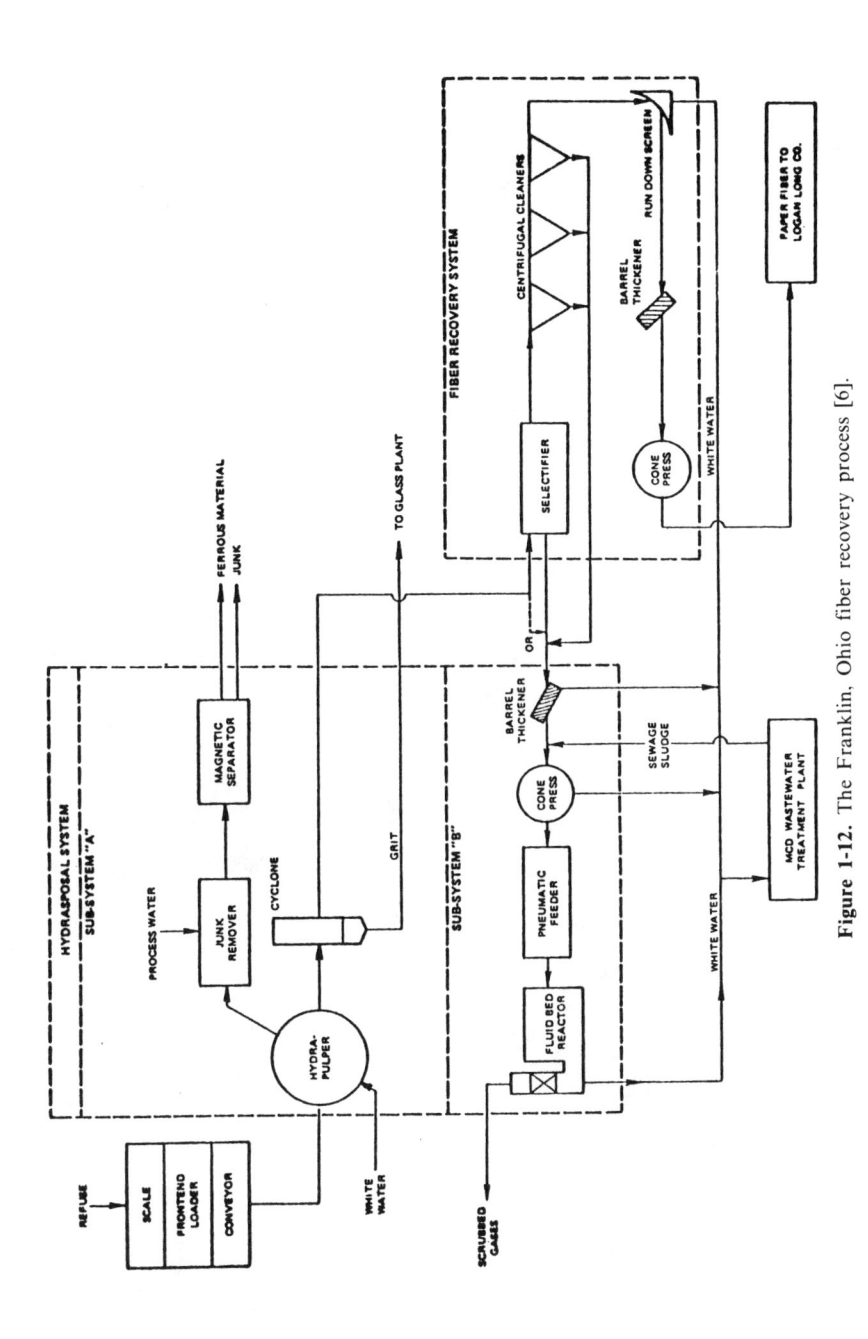

Figure 1-12. The Franklin, Ohio fiber recovery process [6].

20 DESIGN PRINCIPLES

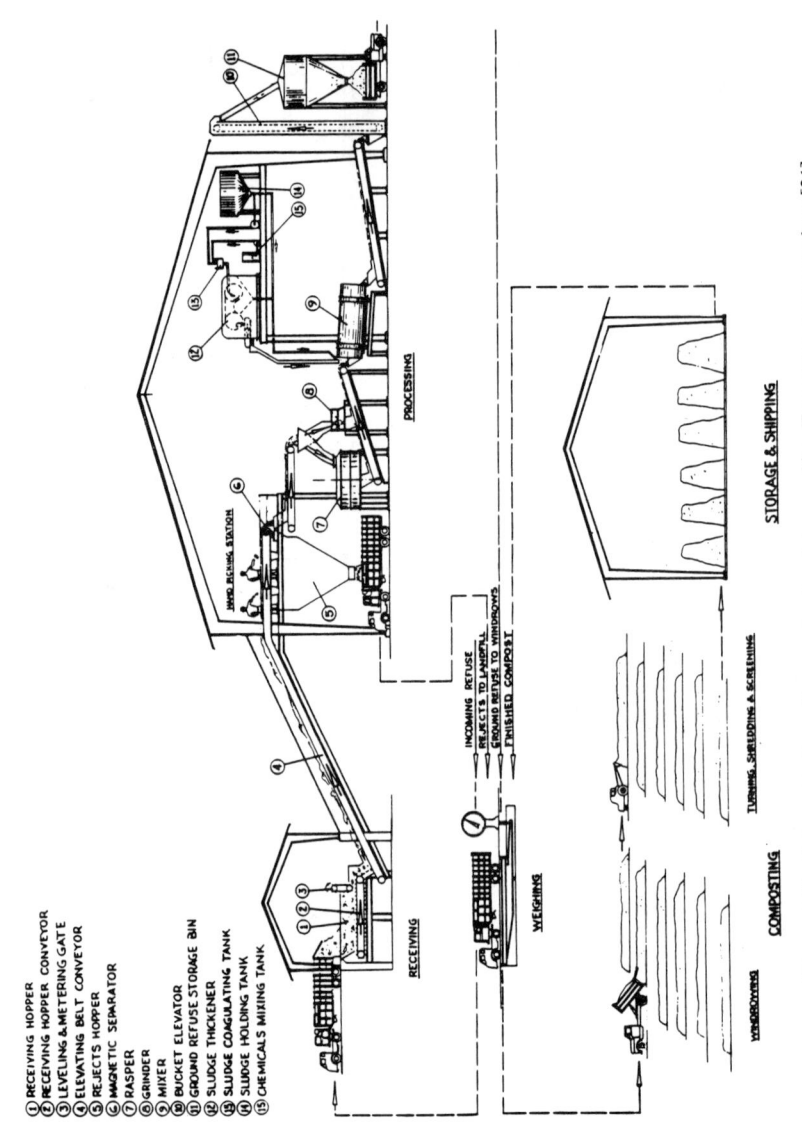

Figure 1-13. Schematic flow diagram for the Johnson City, Tennessee compost plant [21].

Although composting processes (aerobic digestion) have been somewhat successful in Europe, they have not fared well in the United States. A flow plan for a typical compost facility is shown in Figure 1-13 [21].

In addition, a variety of biological and chemical processes are being evaluated in the laboratory for converting the organic portion of the refuse into other types of useful products. Wet oxidation, hydrolysis, fermentation and photodegradation processes have been developed for producing protein, glucose, alcohol and acids from solid waste. However, none of these processes has progressed to a *full-scale* demonstration.

1.6 SUMMARY

In summary, it should be recognized that resource recovery is an emerging technology consisting of a diversity of systems that are experiencing varying degrees of success. To date the most effective systems are those that recover energy from municipal waste. In the overview presented, the major resource recovery systems have been described. A clear knowledge of the different systems is necessary for a better understanding of the design requirements for a resource recovery project. Detailed descriptions of the individual unit processes are presented in Chapter 3.

CHAPTER 2

MUNICIPAL SOLID WASTE

2.1 INTRODUCTION

All sectors of our society generate large quantities of waste. These wastes are produced in a variety of sizes, shapes and compositions and can be looked on as the raw material for resource recovery. Family dwellings, schools, stores, industrial concerns, farms, food processors, etc., all generate glass, metal, paper, plastics and other materials that need to be disposed of safely and effectively. The coverage in this chapter will be confined primarily to the solid waste generated by the municipal sector and some of the nonhazardous solid waste generated by the industrial sector. A comprehensive understanding of the composition, physical, chemical and thermal characteristics, and the fluctuating nature of the waste stream is necessry for effective design of resource recovery facilities. In this chapter, municipal waste composition and characterization are discussed in detail.

2.2 SOURCES OF MUNICIPAL SOLID WASTE

Municipal refuse is a heterogeneous mixture of organic and inorganic wastes discarded by homes, businesses, schools, hospitals and a variety of other entities in the community. A listing of the major sources generating solid waste is presented in Table 2-1. In addition to the municipal sources shown in Table 2-1, solid waste from some industrial facilities (office, cafeteria, packaging and some nontoxic processing wastes, etc.) and small quantities of agricultural wastes (feedlot wastes, truck farm wastes, etc.) also are sent to municipal disposal facilities. All these sources dispose of a

24 DESIGN PRINCIPLES

Table 2-1. Sources of Municipal Solid Waste

Domestic	*Institutional*
Single and multiple dwellings	Schools
	Hospitals
Commercial	Municipal buildings
Offices	
Retail stores	*Municipal Services*
Entertainment centers	Demolition and construction
Restaurants	Street and alley cleaning
Hotels and motels	Landscaping
Service stations	Catch basin cleaning
	Parks and beaches
	Waste treatment residues

variety of materials that may have to be effectively handled in a resource recovery system selected for the community.

2.3 CHARACTERIZATION OF COMPONENTS IN MUNICIPAL SOLID WASTE

As mentioned in Section 2.2, municipal solid waste is composed of a variety of components. A number of different categories have been developed to classify these wastes, the basic ones being paper, glass, metal, plastics, rubber, leather, textiles, wood, miscellaneous materials, and food and garden wastes. A listing of the common materials found in each category is presented in Table 2-2. The quantities of material found in each category will vary considerably and depend to some extent on the geographic location and the season. Although composition varies with time and location, general compositions have been developed to describe MSW in the United States. A generalized composition is presented in Table 2-3. It is very important to remember that the data presented in Table 2-3 are general and that the actual composition will vary considerably, depending on the particular site and time of year. In designing a resource recovery facility for a community, a composition analysis must be conducted several times during each season of the year to obtain an accurate evaluation. During the spring and summer months, yard wastes are at their highest level. During the winter, just the reverse tends to be true (yard waste close to zero). During the spring, all components tend to be closest to their average values, while during the fall all waste components except yard wastes tend to be higher than their average values. Yard wastes in the fall are about half their average value. These seasonal variations will be somewhat different for the southern states,

Table 2-2. Constituents Found in Municipal Wastes [22]

Category	Materials Found
Paper	Newspapers, books, magazines, officer paper, tissues and towels, plates, cups, corregated and other paperboard, paper packaging
Glass	Bottles, jars, light bulbs, glassware and crockery
Metal	Cans, wire, small appliances, barrels, drums, pails, nails, tools, kitchen ware, foil, furniture
Plastics	Packaging, containers (bottles, boxes, bags, etc.), synthetic fabrics, kitchen ware, toys, furniture
Leather and Rubber	Shoes, tires, toys, apparel, luggage
Textiles	Apparel, linen
Wood	Packaging, furniture, toys
Food and Yard Wastes	Garbage (rinds, bones, coffee grinds, leftover food, food trimmings, etc.) plants, leaves, brush, twigs, logs, shrub trimmings, grass
Miscellaneous	Ash, dust and dirt, stones

Table 2-3. Average Municipal Solid Waste Composition as Received

Components	Average Weight (%)	Range (wt %)
Paper	38.0	25–60
Plastics	3.0	1/2–5
Metal	10.0	5–15
Glass	10.0	5–15
Textiles	1.5	1/2–5
Rubber and Leathers	2.5	1/2–5
Wood	3.5	1/7-7-1/2
Food Wastes	16.0	5–35
Yard Wastes	14.0	5–35
Miscellaneous	1.5	1/2–5

where the weather shows less variation between seasons. It should be recognized too that moisture, heat content and other characteristics also will be subject to these variations. In addition to the changes due to season and location, the composition of the waste stream will change with time. As lifestyles change, new technologies become available. Resource availability changes and, as economic conditions change, the waste stream will change to reflect these alterations in daily life. A resource recovery plant usually is designed for a minimum service life of 20 years and must be capable of handling the changing composition of the waste stream. Between 1980 and 2000, Niessen and Alsobrook [22] project increasing percentages of paper, plastics and textile wastes and decreasing percentages of food, yard, metal, glass, wood, leather and rubber wastes. Forecasts by Doggett et al. [23] also project increasing percentages of plastics and paper in the waste stream. To effectively process municipal solid waste, it is apparent that the design parameters need sufficient flexibility to accommodate the changing composition of the waste stream.

2.4 QUANTITIES AND GENERATION RATES FOR MUNICIPAL SOLID WASTE

The size of a resource recovery plant will be dictated by the size of the community it will serve and the quantity of waste generated in it. A common problem encountered in a number of recovery projects has been system overdesign. Unrealistic expectations of available waste have resulted in unnecessary capital expenses in the design of plants to process more refuse than is generated by the community. The quantity of municipal solid waste produced per capita is estimated to range from 3 to 5 lb/day. Based on waste collection data, a generation rate of 3.3 lb per person per day is a reasonable value to use for determining average annual quantities. On average, refuse collected from the residential sector accounts for 2.2 lb; refuse collected from the commercial and institutional for 0.8 lb; and refuse collected from street cleaning and other municipal cleaning and maintenance operations for about 0.5 lb. Although the per capita generation rate has fluctuated over the years, it has tended to be between 3 and 4 lb per person per day. If our affluent lifestyle continues and prosperous economic conditions prevail, it is predicted that the per capita generation rate will increase during the next 20 years [24].

The per capita waste generation rate is based on annual values rather than on a daily or monthly constant. From day to day and month to month, the quantity of waste generated will be subject to considerable variation. Waste generation is usually at its lowest levels during the winter months and at its highest levels during the spring months. The quantity of waste generated declines slightly during the summer and even more during the fall. This seasonal variation has been plotted during two 1-year periods in Montgomery County, Ohio and is presented in Figure 2-1. It can be seen from Figure 2-1 that the quantity of waste generated in a month can vary as much as ±25% of the median value. This large variation can be a problem if not prepared for in the design plan of a proposed waste processing facility.

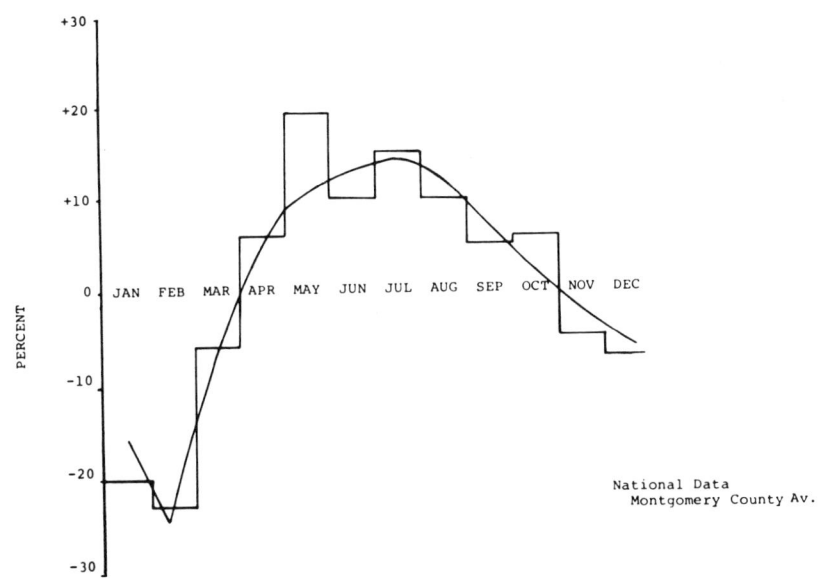

Figure 2-1. Waste delivery to Montgomery County (data provided by Montgomery County Sanitary Engineering Dept.).

2.5 CHEMICAL CHARACTERISTICS OF MUNICIPAL SOLID WASTE

Although the composition of municipal solid is usually defined by the 10 component categories described in Section 2.3, it can also be characterized by its chemical composition. From 60 to 80% of the municipal waste will be classified as organic and will be predominantly cellulose. The organic fraction is provided by the paper, yard, food, plastics, wood, textile, rubber and leather wastes. The inorganic wastes are the metals, glass, pottery, rock and sand. Further, inert fillers in the organics also are classified as inorganic. The inorganic fraction is noncombustible and, after combustion of the waste, constitutes the ash residue. Both the organic and inorganic fractions also contain considerable quantities of moisture. The amount of moisture will vary from 15 to 55% and average about 28%. Food and yard waste contain the larger quantities of moisture (50–75%).

Based on data published from a variety of waste analyses, a summary of the chemical characterizations is presented in Table 2-4, which includes elemental and proximate analyses. From the table it can be seen that municipal waste is an aggregate mixture of waste materials that can be classified into an organic fraction, an inorganic fraction and moisture

28 DESIGN PRINCIPLES

Table 2-4. Chemical Analysis

Components	Wt %	Moisture (%)	Inorganic (%)	Organic (%)
Paper	38	8.0	3.0	27.0
Wood	3.5	0.5	0.3	2.7
Textiles	1.5	0.5	—	1.0
Yard wastes	14	6.0	0.7	7.3
Food wastes	16	10.0	0.9	5.1
Rubber and Leather	2.5	0.5	0.4	1.6
Plastics	3	0.4	0.3	2.3
Metal	10	1.0	9	—
Glass	10	1.0	9	—
Miscellaneous	1.5	0.1	1.4	—
TOTAL	100	28	25	47

Chemical Analysis Element (%)		Organic Composition (%)		Proximate Analysis Component (%)	
Carbon	23.4	Cellulose	79	Volatile matter	42
Hydrogen	3.0	Fat, waxes,		Fixed carbon	5
Oxygen	20.0	oils,		Moisture	28
Nitrogen	0.3	starches,	13	Inert	25
Sulfur	0.1	proteins,			
					100
Other	0.2	rubber,			
Inert	25.0	leather,	3		
Water	28.0	plastics	5		
	100.0		100		

content. The organic fraction, which is almost half the waste, is primarily cellulose (wood fiber) and is of major interest for recovery.

2.6 PHYSICAL CHARACTERISTICS OF MUNICIPAL SOLID WASTE

The important physical characteristics of MSW for the design of waste processing systems are size, shape, density and compaction, as well as the distribution of these parameters in the waste. As municipal solid waste is a heterogeneous mass of constantly changing composition, these physical characteristics have to be viewed on both the aggregate and individual component bases. In addition, the various waste processes will alter the physical character of the waste and require recharacterization after each processing stage. The effect of refuse shredding is of particular importance because it homogenizes the waste, alters the shape of the waste components and reduces the size distribution. General values for bulk density and size distribution for raw and processed refuse are provided to serve as a basis for preliminary design planning.

The density of municipal solid waste is quite variable and depends on the time of year, geographic location, type of refuse collection vehicles, etc. It will represent the composite value of the components present in the waste stream. The bulk density of raw refuse can vary from 150 to 550 lb/yd³ and will average about 220 lb/yd³ (8 lb/ft³). As the moisture content of the waste increases, the bulk density also will increase. The bulk density of wet refuse can range from 350 to 600 lb/yd³. The bulk density of the individual components in waste will be quite varied and depend on the specific density and the disposal configuration (broken bottles, crushed cans, torn paper, etc.). A compilation of the average bulk* and specific densities for major components in municipal solid waste is presented in Table 2-5.

The bulk density of shredded refuse will range from 5 to 20 lb/ft³ and will average about 10 lb/ft³ (270 lb/yd³). The bulk density values for both the shredded and the unprocessed refuse are in the loose condition. Waste

Table 2-5. Average Bulk and Specific Density of Major Components in Municipal Solid Waste

Component	Specific Density (lb/ft³)	Bulk Density (lb/ft³)
Paper	60	7
Plastics	75	2
Wood	40	7
Aluminum	167	—
Iron	491	6
Rubber and Leather	90	15
Food and Yard Waste	45	22
Glass	125	26
Mixed Refuse	—	8
Shredded Refuse	—	10

---

*Bulk density is the ratio of the weight of a material to its overall volume (including any inherent density).

30 DESIGN PRINCIPLES

compacted during storage or for transportation will have a higher density. Waste deposited in a storage pit or pile will compact with time and also have a higher bulk density. To facilitate transportation, hydraulic compaction equipment is used to densify the refuse mechanically. Depending on the pressure applied, the bulk density of refuse can be increased to a maximum of 90–100 lb/ft^3. However, after the compaction pressure is released the compressed refuse will expand to a lower bulk density of 50–60 lb/ft^3, although, as expected, the ultimate density of refuse will be subject to considerable variation. Allowing for moisture and other effects, it can be concluded that refuse can be compressed to a maximum ratio of 6 to 8:1 [25].

The size distribution and shape characteristics of the refuse will vary with the composition of the waste. The size range will extend from grains of sand and dirt to the large bulky items of furniture and household appliances. The morphology of these components also will vary greatly; however, most waste components will be either cylindrical, spherical or platelike in shape. Although there is no definitive description of the morphology of refuse, it is important to recognize that refuse is composed of a diversity of components with a wide range of shapes, which can effect the performance of processing equipment.

Several studies investigating the size distribution and morphology of raw refuse are reported in the literature. The most extensive appears to be the doctural dissertation by Ruf [26]. The cumulative and frequency distribution curves (Figures 2-2 and 2-3) developed by Ruf for the size of components in raw refuse provide a fairly comprehensive evaluation of

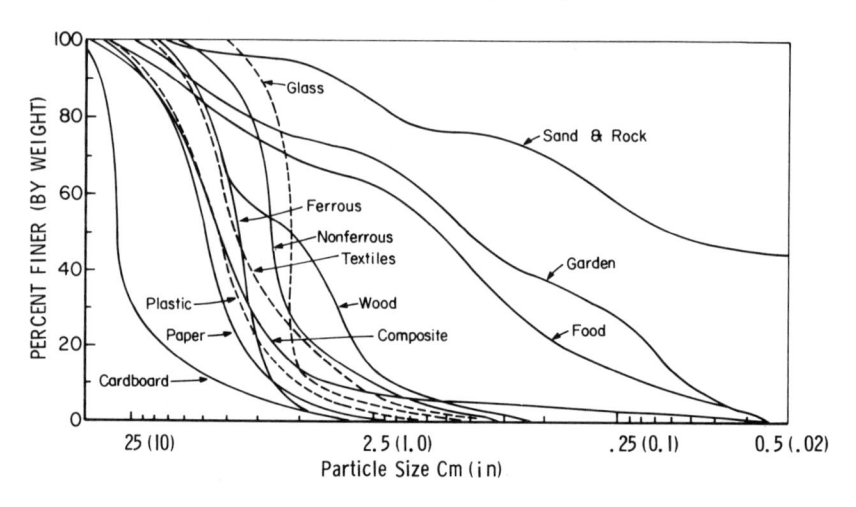

Figure 2-2. Cumulative distributions of raw waste [26].

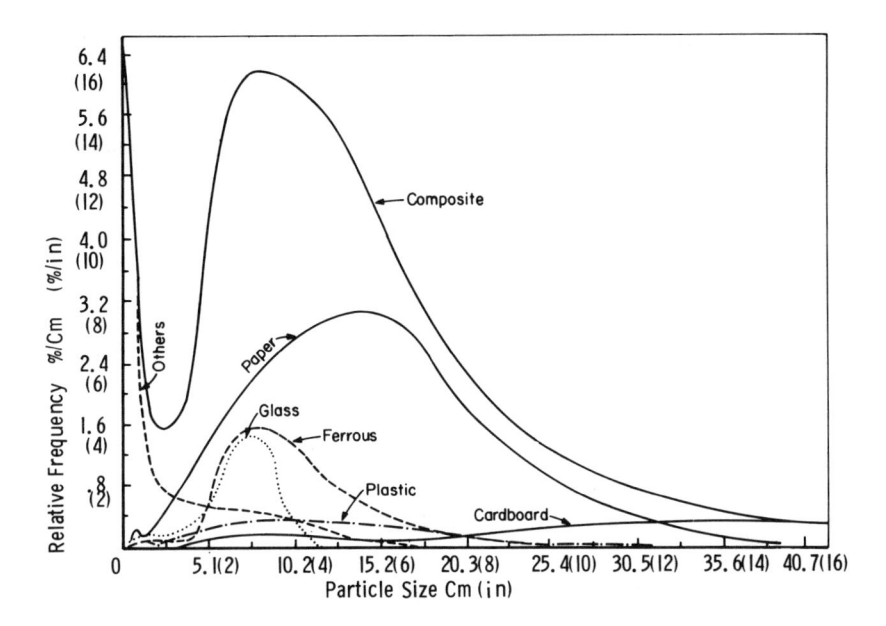

Figure 2-2. Particle size distributions of raw refuse components [26].

size distribution. From the data in these curves, which appear to be in good agreement with other published data, it is possible to determine the average component distribution split of raw refuse. These curves can be utilized to describe the size characteristics of the refuse and serve as a basis for separation processes. From these curves it is possible to predict the recovery distribution for different screen size openings.

Similarly, in his thesis Ruf also investigated the size distribution of primary shredded waste components. The cumulative and frequency distribution curves obtained by Ruf for the range of components in primary shredded waste are shown in Figures 2-4 and 2-5. These data also appear to be in agreement with the other published data. These curves provide a good characterization for primary shredded refuse and also can be utilized to predict separation process performance.

2.7 COMBUSTION CHARACTERISTICS OF MUNICIPAL SOLID WASTE

As a major objective of many of the recovery processes is to utilize the thermal energy in the waste, the combustion characteristics of the refuse are of considerable importance. The heat content is a major measure of

32 DESIGN PRINCIPLES

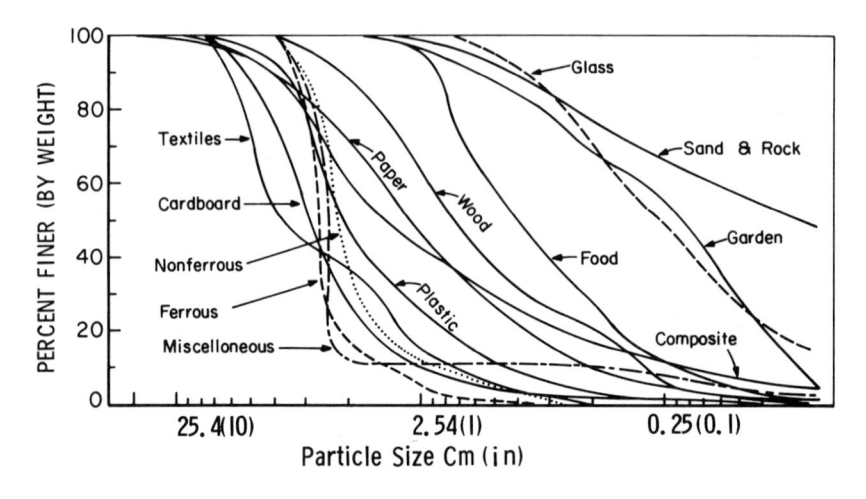

Figure 2-4. Cumulative distributions of shredded waste [26].

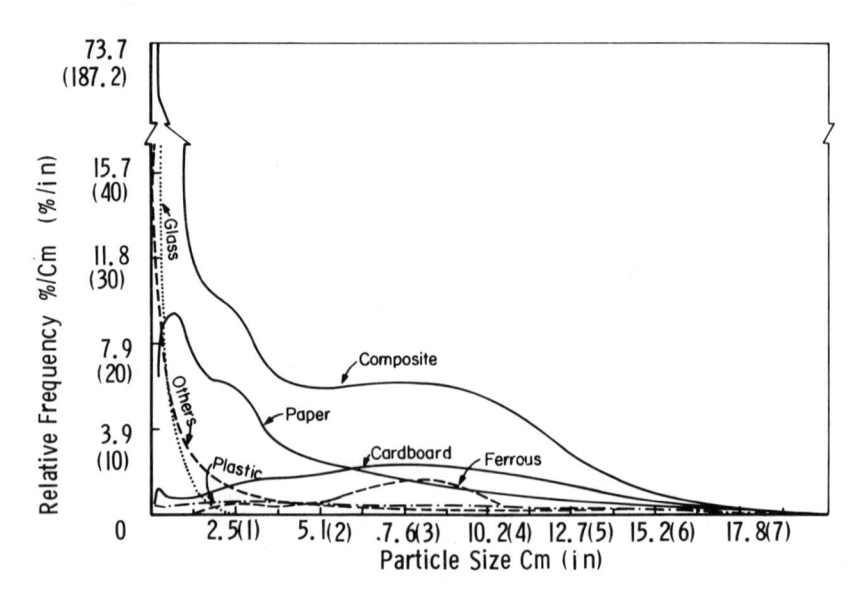

Figure 2-5. Particle size distributions of shredded waste [26].

the fuel value of the refuse. The as-received refuse (28% moisture, 25% inerts) will have an average heat content of 4200 Btu/lb. A reduction in the moisture and inert content will raise the heat content of the waste. Increased quantities of plastics and paper also will increase the heat content. The heat content (or higher heating value) of the as-received refuse can vary from 3500 Btu/lb to 5500 Btu/lb.

A number of processes have been developed for improving the heat content or fuel value of the refuse. RDF (shredded, with most glass and metal removed) will have heat contents that will range from 4500 to 6500 Btu. The powdered RDF (embrittled and pulverized refuse) will have an even higher heat content (~7000 Btu).

2.8 INDUSTRIAL AND AGRICULTURAL WASTES

Although we are concerned primarily with municipal refuse, small quantities of industrial and agricultural waste most likely will have to be processed in municipal resource recovery systems. Niessen et al. [22] have characterized the industrial waste by industry type. A copy of the composition analysis for each industry type developed by Niessen is presented in Table 2-6. The data presented in Table 2-6 can serve as a guide for estimating waste stream composition if local industry is sending, or considering sending, its waste to the municipal system.

The diversity of agricultural processes produces a variety of solid waste. Most of these agricultural wastes do not enter into the municipal waste stream. A brief tabulation of the components in agricultural waste is presented in Table 2-7. Under most circumstances, agricultural wastes will not be a significant factor in the design of resource recovery plants for municipal solid waste.

2.9 MUNICPAL SLUDGES

Municipal sludges consist of a mixture of organic and inorganic phases, suspended in an aqueous solution. Sludges are, predominantly, the solids that settle from the water and wastewater, and/or the colloidal particles that are precipitated by biological flocculation and chemical coagulation. Sludge has a very high moisture content, ranging from 90 to 99%.

The characteristics of the sludge generated from water and wastewater treatment facilities are directly related to the source of the water and wastewater, the type of treatment processes to which the water, wastewater and the sludge are subjected, the type of sewage treatment system and the degree of treatment employed.

In a number of cases, communities have required incorporation of municipal sludge into the resource recovery facility, particularly for mass burning and digestion systems. The codisposal concept is receiving much greater recognition in municipal planning.

34 DESIGN PRINCIPLES

Table 2-6. Industrial Waste Composition [22][a]

SIC	Paper	Wood	Leather	Rubber	Plastics	Metals	Glass	Textiles	Food	Miscellaneous	Tons/ Employee/Year
35											
Data Points	48	48	48	48	48	48	48	48	48	48	48
Average	43.1	11.4	—	—	2.5	23.7	—	0	—	—	3.189
Standard Deviation	34.3	19.5	—	—	6.8	30.8	—	0	—	—	1.438
Confidence Limits[b]	9.7	5.5			1.9	8.7					0.39
36											
Data Points	19	19	19	19	19	19	19	19	19	19	19
Average	73.3	8.3	0	—	3.5	2.3	—	0	1.2	—	2.941
Standard Deviation	24.4	10.1	0	—	7.0	3.5	—	0	2.4	—	7.009
Confidence Limits[b]	11.0	4.5			3.1	1.6			1.1		3.2
37											
Data Points	8	8	8	8	8	8	8	8	8	8	8
Average	50.9	9.4	0	1.4	2.1	—	—	0	—	19.5	2.562
Standard Deviation	34.2	6.3	0	1.5	2.9	—	—	0	—	33.3	4.097
Confidence Limits[b]	23.8	4.4		1.0	2.0					23.1	2.84
38											
Data Points	8	8	8	8	8	8	8	8	8	8	8
Average	44.8	2.3	0	0	6.0	8.4	—	0	—	—	1.769
Standard Deviation	34.0	3.6	0	0	6.4	17.2	—	0	—	—	2.061
Confidence Limits[b]	23.6	2.5			4.4	11.9					1.43
39											
Data Points	20	20	20	20	20	20	20	20	20	20	20
Average	54.6	13.0	—	—	11.9	5.0	—	—	—	8.1	1.603
Standard Deviation	38.7	23.7	—	—	22.2	10.3	—	—	—	14.0	1.901
Confidence Limits[b]	17.0	10.4			9.7	4.5				6.1	0.82

28										
Data Points	48	48	48	48	48	48	48	48	48	48
Average	55.0	4.5	—	—	9.3	7.2	2.2	—	—	19.7	8.862
Standard Deviation	34.0	6.2	—	—	17.0	13.9	4.2	—	—	32.8	10.999
Confidence Limits[b]	9.6	1.7	—	—	4.8	3.9	1.2	—	—	9.3	3.09
29										
Data Points	5	5	5	5	5	5	5	5	5	5	5
Average	72.1	6.8	0	0	15.3	4.4	0	0	0	1.0	1.594
Standard Deviation	35.7	4.4	0	0	30.7	5.2	0	0	0	1.3	2.751
Confidence Limits[b]	31.4	3.9	0	0	27.0	4.6	0	0	0	1.1	2.41
30										
Data Points	13	13	13	13	13	13	13	13	13	13	13
Average	56.3	5.2	0	9.2	13.5	—	0	—	—	—	9.835
Standard Deviation	31.5	6.2	0	20.3	20.7	—	0	—	—	—	9.163
Confidence Limits[b]	17.2	3.4	0	11.0	11.3	—	0	—	—	—	4.97
31										
Data Points	3	3	3	3	3	3	3	3	3	3	3
Average	6.0	3.9	53.3	—	—	13.5	—	0	0	—	8.989
Standard Deviation	4.2	5.4	47.3	—	—	19.2	—	0	0	—	6.986
Confidence Limits[b]	4.7	6.1	53.6	—	—	21.7	—	0	0	—	7.89
32										
Data Points	16	16	16	16	16	16	16	16	16	16	16
Average	33.8	4.3	0	—	—	8.1	12.8	—	0	40.0	6.412
Standard Deviation	37.5	8.4	0	—	—	24.8	29.6	—	0	44.8	15.300
Confidence Limits[b]	18.4	4.1	0	—	—	12.2	14.5	—	0	22.0	7.48
33										
Data Points	12	12	12	12	12	12	12	12	12	12	12
Average	41.0	11.6	0	—	5.4	5.5	2.0	0	—	29.0	3.184

MUNICIPAL SOLID WASTE 35

36 DESIGN PRINCIPLES

Table 2-6. Continued

SIC	Paper	Wood	Leather	Rubber	Plastics	Metals	Glass	Textiles	Food	Miscellaneous	Tons/Employee/Year
Standard Deviation	27.4	12.4	0	—	9.8	7.8	4.3	0	—	40.0	15.796
Confidence Limits[b]	15.5	7.0			5.5	4.4	2.4			22.7	8.93
34											
Data Points	36	36	36	36	36	36	36	36	36	36	36
Average	44.6	10.3	0	—	—	23.2	—	—	—	12.2	6.832
Standard Deviation	37.7	20.8	0	—	—	34.5	—	—	—	31.0	9.088
Confidence Limits[b]	12.3	6.8				11.3				10.1	2.95
20											
Data Points	30	30	30	30	30	30	30	30	30	30	30
Average	52.3	7.7	—	—	0.9	8.2	4.9	0	16.7	9.2	7.949
Standard Deviation	32.7	10.9	—	—	0.4	3.7	2.8	0	29.9	21.1	8.760
Confidence Limits[b]	11.7	3.9			0.1	1.3	1.0		10.7	7.5	3.13
22											
Data Points	18	18	18	18	18	18	18	18	18	18	18
Average	45.5	—	0	—	4.7	—	—	26.8	—	—	2.160
Standard Deviation	40.3	—	0	—	10.7	—	—	38.1	—	—	1.900
Confidence Limits[b]	18.6				4.9			17.6			0.86
23											
Data Points	17	17	17	17	17	17	17	17	17	17	17
Average	55.9	—	0	0	—	0	0	37.3	2.8	—	6.211
Standard Deviation	37.4	—	0	0	—	0	0	37.3	2.8	—	6.211
Confidence Limits[b]	17.8							17.7	1.3		2.94

MUNICIPAL SOLID WASTE 37

	Data Points	Average	Standard Deviation[b]	Confidence Limits[b]
24	9	16.7	33.6	22.0
	9	71.6	34.8	22.7
	9	0	0	0
	9	0	0	0
	9	—	—	—
	9	—	—	—
	9	0	0	0
	9	0	0	0
	9	7.8	19.7	12.9
	9	8.531	7.648	4.97
25	7	24.7	12.3	9.1
	7	42.1	16.2	12.0
	7	0	0	0
	7	—	—	—
	7	—	—	—
	7	0	0	0
	7	—	—	—
	7	—	—	—
	7	—	—	—
	7	2.783	3.578	2.64
26	20	56.3	8.7	3.8
	20	11.3	15.5	6.8
	20	0	0	0
	20	—	—	—
	20	9.4	18.2	8.0
	20	—	—	—
	20	—	—	—
	20	—	—	—
	20	14.0	27.5	12.1
	20	3.987	8.330	3.64
27	26	84.9	5.8	2.2
	26	5.5	12.3	4.7
	26	—	—	—
	26	0	0	0
	26	—	—	—
	26	0	0	0
	26	—	—	—
	26	—	—	—
	26	—	—	—
	26	5.835	12.575	4.82

[a] Courtesy of Arthur D. Little, Inc.
[b] 95% confidence limits.

Table 2-7. Tabulation of Agricultural Solid Waste Components

Manure
Forestry Residues
Crop Residues
Crop Spoilage
Pen Sweepings
Food Processing Wastes
Dead Animals

2.10 SUMMARY

Municipal solid waste is a heterogeneous mixture of metal, glass and organic materials discarded by the residential, commercial and institutional sectors of the community. As received, the refuse contains about 28% moisture, 10% glass, 10% metal and 52% organic materials (with about 5% inert filler). The refuse has an average bulk density of 8 lb/ft^3 and an average heat content of 4200 Btu/lb. Although both average and specific property values are available for municipal solid waste, it is important to recognize that both composition and generation rates are fluctuating constantly. Hourly, daily and seasonal changes typify the character of municipal solid waste.

CHAPTER 3

BASIC PROCESSING TECHNOLOGIES

3.1 INTRODUCTION

The diversity of resource recovery systems developed rely on a multitude of basic material processing technologies:

- Size reduction (comminution)
- Separation
- Handling (transport and storage)
- Thermal processes (drying, combustion, etc.)
- Biological and chemical processes

This chapter describes these basic unit operations as they apply to municipal solid waste processing and are applied to resource recovery systems. Both basic theory and applied technology are discussed.

3.2 MATERIAL SIZE REDUCTION

Size reduction and material homogeneity is a major processing step for a number of recovery systems. A wide range of equipment is available for reducing the size of refuse to facilitate more efficient separation and recovery processes. Tension, compression and shear forces are employed that impact, crush, grind, tear, cut, pulverize and shred the refuse. Shredding is the general term applied to these mechanical processes that reduce the size and homogenize the refuse. Most processes utilize either a grinding or impacting action to reduce the refuse; however, both tearing and shredding are likely to occur in combination with the grinding or impacting.

40 DESIGN PRINCIPLES

The eleven basic types of size reduction equipment are crushers, cage disintegrators, shears, shredders, grinders, cutters and chippers, raspe mills, drum pulverizers, disc mills, wet pulpers and hammermills [27]. Hammermills and wet pulpers are the two major types of equipment used for size reduction of refuse.

3.2.1 Hammermills

As stated, hammermills are the most commonly employed equipment used for solid waste reduction. The main element of the hammermill is the central rotor with radial arms or hammers, which protrude from the circumference of the rotor. The rotor is encased in a heavy-duty housing. In some designs, stationary breaker plates or cutter bars are mounted inside the housing unit. Two basic types of hammermills are available on the market: horizontal rotor shaft or vertical rotor shaft. The horizontal rotor shaft unit is the more common type. In both units, raw refuse is introduced from the top and flows through the machine by gravity, exiting from the bottom openings in the unit. In the horizontal-type hammermills, the size of the existing refuse is controlled by the size of the openings in the floor grate. In the vertical shaft hammermill, exiting size is controlled by the space between the lower hammers and the breaker-bars on the inner housing wall. Cross-sectional views of both the vertical- and horizontal-type hammermills are shown in Figures 3-1 and 3-2, respectively. As shown in Figure 3-1, size reduction of raw refuse in the horizontal shaft hammermills is achieved by more of a grinding action (see Figure 3-2) [27,28].

Most hammermills designed for shredding municipal solid waste require a main drive motor in the range of 500 to 1500 hp. These motors draw substantial current and will require separate substations. The size and power required for any particular hammermill will be determined by the nature of the raw refuse, the processing rate and the outward particle size required. In most cases, primary shredding of the raw refuse will reduce all the solid waste to minus 6 in. in size. Processing rates of 50–100 ton/hr can be achieved with these types of hammermill units. The size distribution of the material exiting in the hammermill is quite variable and will be a function of the refuse composition, moisture content, feedrate, size openings in the exist grate and rotor speed. Shredded municipal refuse from five different sites was obtained by the National Center for Resource Recovery (NCRR) for evaluation [29]. A plot of the size distribution data reported is presented in Figure 3-3 for purposes of comparison.

In a hammermill, the metal usually will be sliced and crumbled. Paper

BASIC PROCESSING TECHNOLOGIES 41

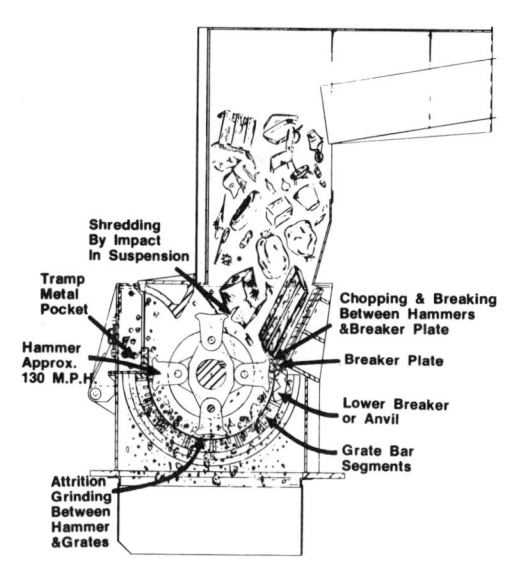

Figure 3-1. Horizontal hammermill (courtesy Gründler Crusher and Pulverizer Co.).

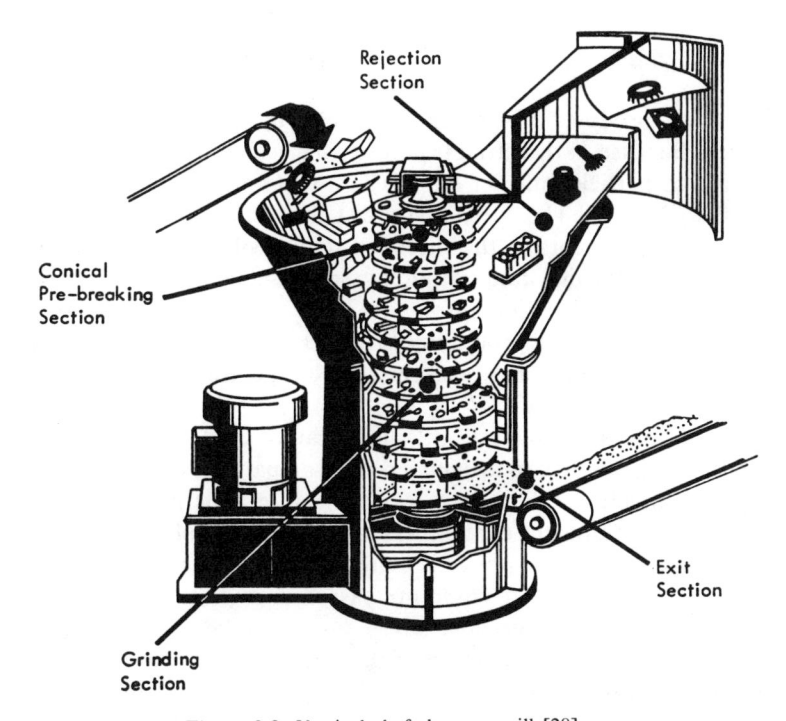

Figure 3-2. Vertical shaft hammermill [28].

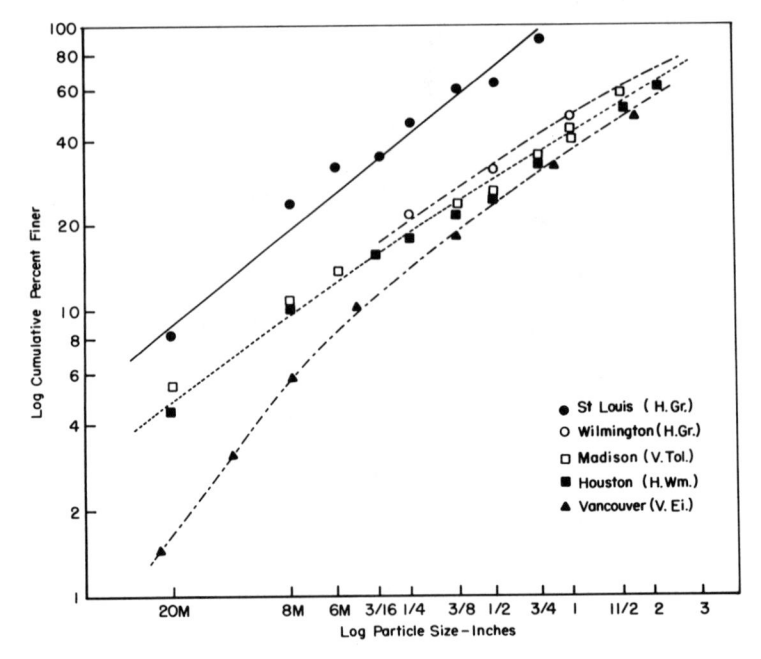

Figure 3-3. Particle size distribution of shredded MSW [29].

and wood products will be shredded and the glass and ceramics pulverized. Becuase of their friable nature, all glass and ceramic materials are reduced to particles smaller than 1/2 in. in diameter and the majority to less than 3/16 in. The specific size distribution and morphology of the different components in the shredded refuse will vary from day to day depending on the moisture content of the refuse, the wear condition of the hammers, the processing rates and input power requirements of the hammermill unit. The processing variables for shredding municipal solid waste in hammermills has been the subject of considerable study [27,30–33].

In addition to hammermills, flailmills can be used for shredding refuse. In a flailmill, two sets of articulated flails on parallel shafts are rotated in opposite directions. The flails tear the refuse as it passes through the rotors. This equipment can process large quantities of refuse at very low power levels. The low power requirements are due to the fact that the items that are difficult to shred are passed through the machine without being significantly reduced in size.

The secondary size reduction of shredded refuse is achieved by the use of hammermills with reduced grate size openings, roll grinders, disc mill grinders and ball mills. The secondary grinding units can reduce the shredded refuse from a nominal 6 in. to a fine powder. Secondary hammermills normally reduce the shredded refuse to a nominal minus 1 in. size. Mechanical grinders can reduce shredded refuse to a powder.

Incorporation of a hammermill into a resource recovery system also requires the use of auxiliary equipment to facilitate the operations of the hammermill. Conveyors are required to feed the hammermill and carry away the discharged material. Dust control systems are needed to minimize health and environmental problems. In addition, fire and explosion control systems are required to protect the health and safety of plant personnel. An overhead crane also is frequently required to facilitate maintenance and repair operations on hammermill units, and control instruments are necessary for the operation of the hammermill and its auxiliary equipment. Incorporation of a hammermill into a resource recovery system also requires an appreciation of some of the problems frequently associated with the operation of size reduction equipment. Overfeeding, jamming, excessive hammer wear, fire and explosion are just a few of the common types of problems encountered by hammermill operators. Explosions are the most difficult problem to deal with in the daily operation of a facility.

Size reduction of municipal refuse also can be achieved by the use of wet processing equipment. The most common type of equipment used is a wet pulper, which can be viewed as an underwater hammermill. The refuse is introduced to the pulper in the form of a slurry made up of 10% solids and 90% water. The central rotor revolving at high speeds forms a vortex in the slurry, and the repeated impacts of solids in the slurry with the rotating element reduce the solid waste to a pulp. The unpulpable items are rejected, and the slurry pulp is carried off for further processing. From the pulper, the slurry is transported to a liquid cyclone for separation of the pulped organic materials from the inorganic solids in the slurry. The light, predominantly fibrous, material then is dewatered for further recovery processing [6].

3.3 SEPARATION

Recovery of the valuable component in refuse can be achieved by various types of separation equipment. These separation operations rely on a diversity of physical, chemical and thermal mechanisms, as well as on the characteristics of the waste fraction selected for separation. Component size, density, magnetic susceptibility, electrical conductivity, optical characteristics, chemical properties and thermal properties are the major means used for separating different fractions in the waste. Size, density and magnetic separation processes are the major types of unit operations used in resource recovery operations.

3.3.1 Size Separation

Due to its heterogeneous nature, both the size distribution and shape characteristics of the refuse are quite varied. The size range will extend from grains of sand and dirt to large bulky items like furniture and home appliances. Accordingly, size separation is one means of segregating refuse components for more effective processing. Screening or size separation equipment is used in processing both raw and shredded refuse at a number of resource recovery facilities. Effective screening processes for municipal solid waste require that a comprehensive characterization of the size and shape distribution of the components in raw and shredded refuse be established. Size and shape distribution data for the components in refuse are described in Chapter 2 (Section 2.6).

A wide range of equipment is available for screening raw and shredded municipal refuse. This equipment can be divided into five main categories: grizzly screens, revolving screens, shaking screens, vibrating screens and oscillating screens. Grizzly screens are used primarily for scalping and consist of a pair of parallel bars separated by a predetermined distance. Grizzly screens either can be stationary or agitated. Shaking and vibrating screens consist of woven wire cloth or perforated punch plate fastened to metal or wooden frames. The frame usually is inclined slightly and is driven with a shaking or vibrating motion. The material to be screened is introduced at the upper end of the frame and is advanced by the motion of the screen while the finer particles pass through the openings. Oscillating screens, similar to vibratory and shaking screens, are box-like round or square devices with a series of screen surfaces fixed one on top of another and oscillated in a circular or near circular orbit. The oscillating screens are characterized by low-speed oscillation in a plane almost parallel to the screen surface. Revolving or trommel screens consist of a rotating cylindrical frame covered with a wire cloth or perforated plate, open at both ends and inclined at a slight angle. These trommel screens have proved relatively effective for a number of the size separation applications required in the processing of municipal waste. In addition to trommels, rotary disc screens also have been employed recently for the screening of shredded refuse. Rotary disc screens can be described as an array of rotating cylinders separated by a fixed distance. The smaller particles in the refuse being carried over the rotating cylinders fall into the openings, while the oversized particles are carried along the top of the rotating cylinders to the discharge chute [34-38] (Figures 3.4-3.7).

Selection of the appropriate processing parameters of the screening system will depend on the condition of the refuse and the requirements for the fraction to be recovered. The technology for trommelling municipal solid waste serves as a good model for demonstrating the role of the different processing parameters important to the design of size separation systems.

The rotary screen or trommel has been an effective processing unit at a number of resource recovery plants. The trommel can be used prior to shredding to reduce the quantity of waste to be shredded and to concentrate the metal and glass fractions for subsequent recovery. Trommelling will break open the trash bags, separate many of the agglomerates and concentrate the glass and metal fractions for easier

Figure 3-4. Grizzly screen [35].

46 DESIGN PRINCIPLES

Figure 3-5. Shaking or vibrating screen [36].

recovery. Further, the oversized fraction processed through the trommel is likely to be a more desirable feed material in a mass burning process. The trommel also can be a useful processing device in the preparation of refuse-derived fuel from shredded refuse. Separating shredded refuse into several size fractions prior to air classification will further enhance component separation. In addition, a trommel can be used to separate the glass and fine fraction from coarse or primary shredded refuse, and thus enhance its value as a fuel.

Trommels are revolving screens consisting of perforated cylindrical tubes mounted on drive units. The trommel is usually mounted on an incline with the feed introduced at the upper end of the tube, and the material is screened as it tumbles down the tube. The undersized material passes through the openings in the cylinder wall, while oversized material exists at the lower end of the trommel. The material tends to follow a helical path through the length of the tube. The openings in the trommel can be of a single size or there can be two or three zones of different size openings. in addition, compound trommels consisting of two or more concentric screening tubes on the same axis can be used for multifraction separation. Trommels with two or more zones of different size openings or compound trommels may be effective in concentrating selected components in municipal solid waste. These units could provide a means for improving the quality of the combustible fraction, or a glass- or metal-rich fraction.

The design specification and operating parameters for trommels are dictated by the physical characteristics of the material to be screened, the required feedrate and the separation requirements. The primary design parameters include trommel length, diameter, slope and aperture size.

BASIC PROCESSING TECHNOLOGIES 47

Figure 3-6. Trommel [37].

The primary operating parameters are rotation velocity and feedrate. The length of the trommel selected will determine the retention time of the material in the unit and affect the screening efficiency. The greater the trommel length, the more complete will be the removal of undersized material, although the majority of screening occurs in the first few feet of the unit. However, for screening raw MSW, it must be remembered that a considerable portion of the material will be in plastic or paper sacks and that the sacks will have to be broken open in the first few feet of travel. Thus, in the initial few feet of processing only a fraction of the undersized material will come into contact with the openings in the wall of the trommel. For most mineral ore dressing applications, trommel length rarely exceeds 10–15 ft. For screening MSW, trommel length selection will be based on the nature of the material and the separation efficiency desired. To date this selection has been based mostly on the trial and error method.

The diameter selected for a trommel unit will determine in large part the capacity of the unit and the thickness of the bed during processing. For a fixed feedrate, a larger trommel diameter will result in a thinner bed of refuse. One correlation developed for determining trommel diameter for given capacities is described by Taggart [34] where D (the diameter in inches) is equal to $7.66\sqrt{C/Gs}$, C is capacity in short tons per hour, and Gs is a specific gravity of the feed material. For MSW, specific gravity will range from 1 to 1.5. It must be recognized, however, that this formula was developed for the screening of coal, sand and stone-type materials and may not be appropriate for MSW, which is not as free flowing a feed

48 DESIGN PRINCIPLES

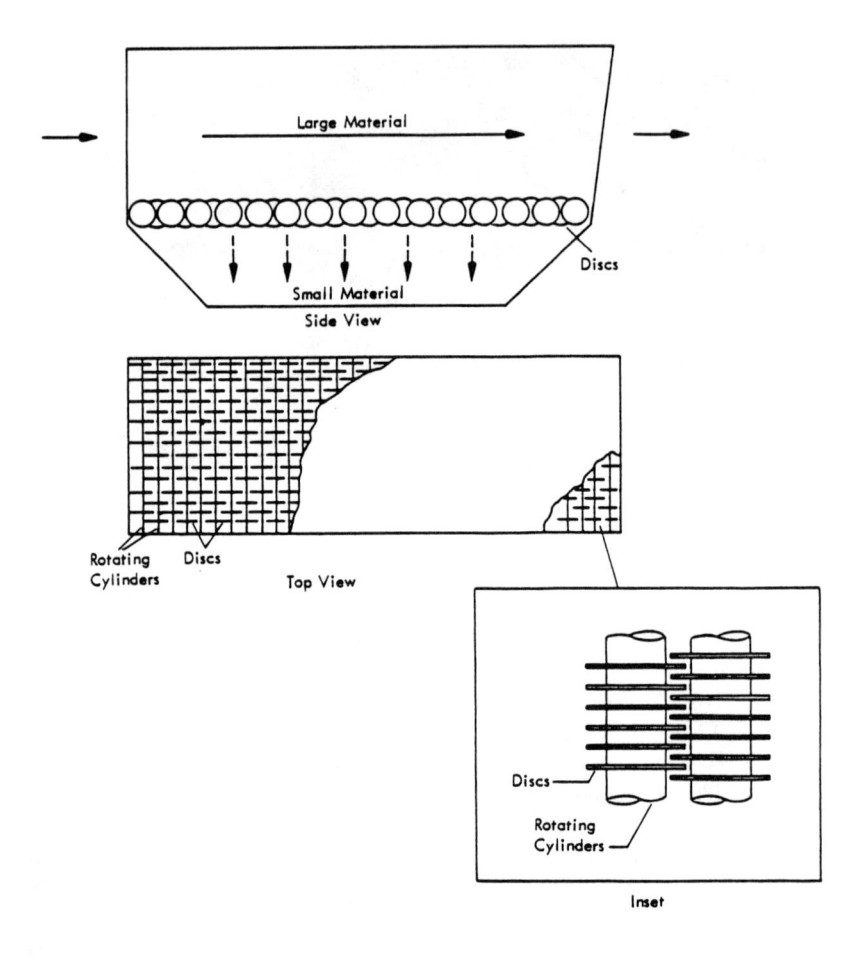

Figure 3-7. Rotating disc screen [38].

material. In addition, the initial process for MSW is the breaking open of the plastic and paper sacks, the breaking apart of agglomerates and the breaking of the large glass and ceramic products. Trommel diameter and rotating speed must be selected carefully to allow the cascading and cataracting actions to open the bagged and compacted material. The tumbling motion given to materials in a rotating shell can be of two kinds:

1. rotation around its own axis, or
2. cascading/cataracting;

BASIC PROCESSING TECHNOLOGIES 49

Cascading is defined as rolling down the surface of the load and cataracting as parabolic free fall above the mass [39].

Based on its experience as a manufacturer of screening equipment, Triple/S Dynamics has developed a different approach for determining the diameter to capacity relationship for trommel units processing MSW [40]. The company's experience has shown that the best results are obtained when the volume of MSW being processed is about 25% of the available volume in the trommel. To calculate the desired diameter, Triple/S Dynamics assumes an average MSW bulk density of 6 lb/ft^3 and an average velocity through the trommel of 15 ft/min. The resultant formula for determining trommel diameter is then $D = 24\sqrt{C/2.12}$. Using the formula presented by Taggart and the approach and assumptions developed by Triple/S Dynamics, the diameter to capacity relationships are compared in Table 3-1.

In New Orleans, a 10-foot-diameter trommel installed for processing MSW has reported a feedrate capacity of more than 100 ton/hr [41]. The data from the New Orleans facility show a capacity of almost twice that calculated by the Triple/S formula, about one-third that calculate by the formula of Taggart. This brief exercise further demonstrates the need for more effective design data for trommels to process MSW.

The slope of the trommel employed will affect the rate at which the material travels through the unit. For a given feedrate, the greater the slope the thinner the bed, thus providing for higher efficiencies of screening. Taggart states that for punch-plate trommels a slope of five degrees is recommended, and for woven-wire units a slightly greater slope is required [34]. In New Orleans, the slope selected for the trommel unit was five degrees [41]. Trommel slopes of three, four and five degrees were evaluated in a study for the National Center for Resource Recovery and the Tennessee Valley Authority (TVA), and little difference was observed [42,43]. At the University of California's Solid Waste Processing Laboratory, Savage et al. [44] and Savage and Trezek [45] used a trommel with an inclination of 15 degrees to separate the fine fraction

Table 3-1. Comparison of Diameter to Capacity Relationship

Diameter		Capacity-ton/hr	
(in.)	(ft)	Taggart	Triple/S
36	3	28	5
48	4	50	9
72	6	110	19
120	10	307	53

from shredded municipal solid waste. For a fixed feedrate, trommel slope and rotation speed have to be selected properly to provide effective screening of the MSW materials.

Further, trommel rotation speed also will affect both the capacity and efficiency of the trommel unit. As rotating speed increases, trommel efficiency will pass through a maximum and then decrease sharply. Maximum efficiency appears to correspond to a rotating speed, which causes the load to ride about two-thirds of the way to the top of the screen. The height to which the refuse will travel along the side of the trommel before it tumbles down (cataracting and cascading) is determined by the critical speed for the particular unit. The critical speed is defined by the equation v_c (in rpm) = 76.6 D, where D is the diameter (in ft) of the trommel, and v_c is the velocity at which centrifugal force on the waste in contact with the shell at the height of its path equals the force on it due to gravity. At speeds greater than v_c, the waste would be carried around the shell of the rotating trommel [34]. Rotating speeds normally will range from 35 to 40% of critical speed. The critical speed for a 3-foot-diameter trommel would be 44.2 rpm; for a 4-foot diameter, 38.3 rpm; and for a trommel with a 6-foot diameter, 31.2 rpm. In recent studies with trommels processing municipal solid waste, rotational speeds of from 9 to 30 rpm were reported [41-45].

In addition to length and diameter variations, trommels can be designed with different configurations—cylindrical, conical or hexagonal. Conical units have been used to a considerable extent in gravel-washing plants, while hexagonal trommel units have been used for the screening of fine materials.

The walls of the trommel can be constructed from wire mesh or can be punched-metal screens. Although punched-plate screens have a longer life and can be prepared with a wide diversity of aperture designs, they have a lower percentage of open area. The opening in the woven-wire screen will be either square or retangular. In the punched-plate screen, the openings could be round (in a straight or staggered array), square, rectangular (in a straight or staggered array) or oblong (in a straight or staggered or diagonal array).

Aperture size, spacing and shape will be dictated by the fraction identified for separation and the required feedrate. Material separation efficiency is a statistical probability function based on the number of collisions of undersized material with the screen openings and the capacity of the bed for dry creening. A number of rules of thumb for estimating screening efficiency versus capacity also are discussed in the literature for mineral processing. The screening efficiency of trommels will vary from 70 to 90%.

3.3.2 Density Separation

Components and groups of components (fractions) in the raw and shredded municipal waste also can be separated by the utilization of density classification equipment. The most commonly employed dry density separation equipment used for the processing of municipal refuse is the air classifier; however, it should be noted that for air classifiers, particle size and shape play a role in the separation process. A number of these classifiers have been developed and employed for separating out the light combustible material from municipal waste by the use of a moving air current. Vibrating tables, jigs and wet classifiers are other examples of density classification equipment that also have been used in the processing of municipal waste. Wet classification units are used to float off the organic fraction for recovery or for the separation of metals from the heavy fraction of municipal refuse. Jigs have been used primarily for recovery of different mineral fractions and metals in the processed waste.

Air classification is the term used to describe a number of processes that utilize gravity and air currents for material separation when the material to be separated is introduced into a stream of moving air. The three main factors that affect the separation of particles in an air stream are particle size, shape and density. Vertical, horizontal and inclined column separator designs have been developed that utilize these parameters to separate the different fractions in solid waste. Component density is the major factor for separation by this gravitational process. A compilation of waste component densities is presented in Table 3-2.

To suspend a particle in an air stream, the drag must equal the weight:

$$\text{Weight} = \text{Drag}$$

$$V_s(\rho_s - \rho_a)g = \tfrac{1}{2} \rho_a C_D V_a^2 S_s$$

Table 3-2. Waste Component Densities [46]

Component	Material Density (lb/ft^3)	
Paper	60	(44–75)
Plastic	75	(56–100)
Wood	40	(19–56)
Aluminum	167	
Iron	491	
Rubber and Leather	90	(62–125)
Food and Yard Wastes	45	
Other Nonferrous Metal	550	
Glass	125	(144–187)

52 DESIGN PRINCIPLES

where: V_s = volume of the material
ρ_s = density of the material
S_s = surface area of the material exposed to the air stream (lift area)
ρ_a = density of the air
C_D = drag coefficient for the material
v_a = air stream velocity
g = gravitational force

If the drag force exerted on any particular particle in the process stream is greater than the gravitational force on that particle, it will be lifted and carried along with the air stream. Conversely, if the gravitational force is greater than the drag force, then the particle will fall from the moving air stream. Thus, a separation is effected between the less dense particles (lights) and the heavier particles (heavies). The drag equation presented above provides an effective insight to the air classification process. As shown, the drag force ($\rho_a C_D V_a^2 S_s$) depends on the square of the air stream velocity. This shows that small changes in the air velocity have a large effect on what is carried in the air stream and, therefore, on the separation efficiency. In addition, it demonstrates that an air classifier will be very sensitive to air velocity and will be difficult to "fine tune" by changing the air velocity to obtain a particular separation ratio. From the drag equation, it is also noted that the drag force is dependent on the area of the particle presented to the air stream. A spherical shape would preset the same area, of course, no matter what its orientation, but an irregularly shaped object will present a wide range of effective areas, depending on its orientation with respect to the air stream. Thus, large flat plates otherwise expected to drop out of the air stream may be carried along with the less dense spherical particles. Simply stated, if two particles of the same size and shape have different densities, they can be separated by an appropriate air velocity, which will lift and carry the less dense particle, while the more dense particle drops out of the air stream. Unfortunately, municipal solid waste is not homogeneous in size, shape or particle density. Furthermore, its composition is not even consistent from site to site or from day to day at any one site. In addition, compaction during collection promotes municipal solid agglomeration, which makes separation of individual components difficult.

It is desirable for effective separation of waste components to break up any agglomerations that may form. This can be accomplished in most air classifier systems by the use of turbulent zones. In a turbulent region, the effective velocity of the air stream is changing both in magnitude and in direction so that the particles are subjected to a rapidly changing drag force. Such zones of turbulence in an air classifier help break up agglomerates into separable constituents, allowing the lights to be carried in the air stream and the heavies to drop out. Furthermore, the turbulent zones also help further separate individual light and heavy particles so that one is not carried along in the wake of another. Turbulent regions can be designed into an air classifier in a number of ways. They can be introduced by forcing the air stream to make an extreme change in direction, by expanding or contracting the size of the pipe enclosing the air stream or by introducing jets of air at an angle to the main air stream velocity. These approaches, or combinations of them, are used in various ways by all manufacturers of air classifier systems presently in use or planned [47-55].

Both horizontal and vertical classifiers are available for the dry separation of refuse. In the vertical units, the upward flow of air in the enclosed column picks up and carries the lower-density materials as it contacts the downward falling refuse. Air stream velocity will dictate the separation ratio (lights versus heavies). This separation is predominantly a function of the density and surface area of the shredded refuse particles. To compensate for the day-to-day variations in particle moisture content and morphology, a variable air velocity control is necessary.

At present, two types of vertical classification units are on the market. One system incorporates a zig-zag design, in which the bulk shredded refuse is introduced into a vertical air stream as it tumbles down the zig-zag baffles. Each zig-zag provides an additional point of decision for the separation because each zig-zag is a point of turbulence in the air stream. The tumbling action at these points causes agglomerated materials to be broken up. The second type of unit is essentially a vertical column, in which the shredded refuse is injected midway into the column, which has an upward air stream. The column can be rectangular or cylindrical and may contain baffles that can be tilted. These vertical column units require more closely sized refuse than do the zig-zag units. Manufacturer specifications require the refuse to be minus $1^{1}/_{3}$ in. in size (Figure 3-8) [56].

Several types of inclined and horizontal air classifiers also have been developed for separating both raw and shredded refuse. In the horizontal unit, the refuse is introduced into a horizontal air stream and the refuse components will drop out of the air stream at different points along the flow path as a function of their weight and shape. In most designs, bins are placed under the drop zones, and the very light material is carried out of the classifier with the air stream. In the horizontal system, two to five separate fractions can be recovered depending on the design. The refuse can be introduced into the classifier by dropping it past a horizontal air stream, or by using a ballistic wheel, which hurls the refuse into the horizontal air stream. Designs employing rotating inclined cylinders also have been developed to classify municipal refuse. In these systems, lifters

54 DESIGN PRINCIPLES

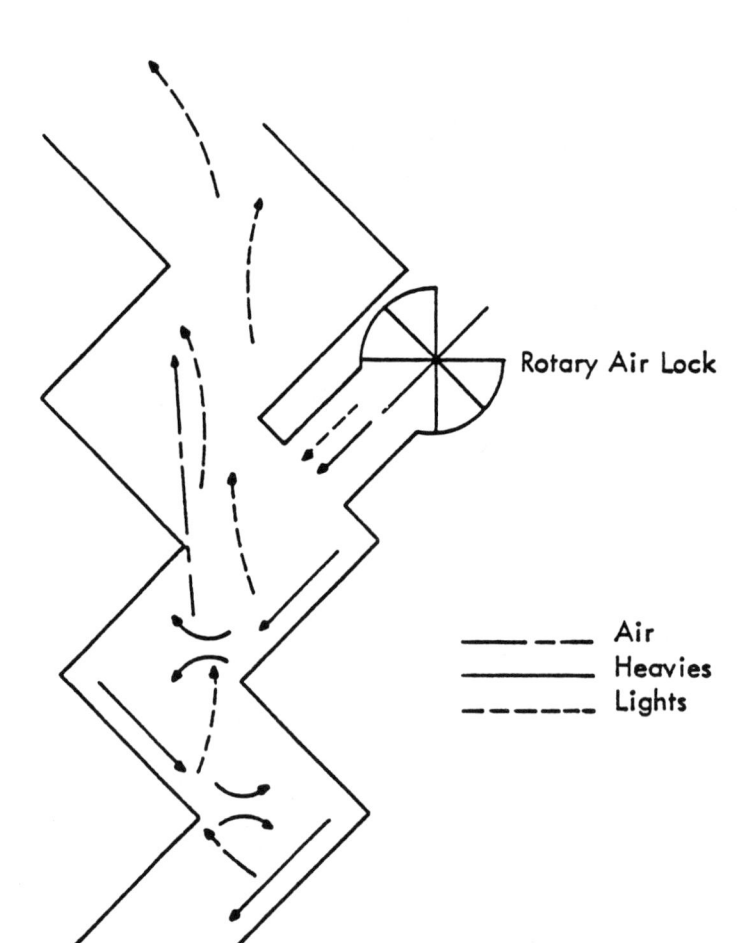

Figure 3-8. Principle of zigzag air classifier [56].

within the cylinder cause the refuse to cascade down the tube as an upward air flow carries the lights out of the unit, while the heavy fraction is discharged out of the bottom of the cylinder (Figure 3-9) [57].

Another type of horizontal unit is a vibrating air classifier based on the modification of a gravity or air table. In a vibrating air classifier, the shredded refuse is conveyed down a series of stepped vibrating pans. At each step air passes through the material carrying off the light particles (Figure 3-10) [58].

The light fraction from the different air classifiers is pneumatically conveyed to a cyclone for deairing. A cyclone is primarily a settling chamber in which gravitational acceleration is replaced by centrifugal

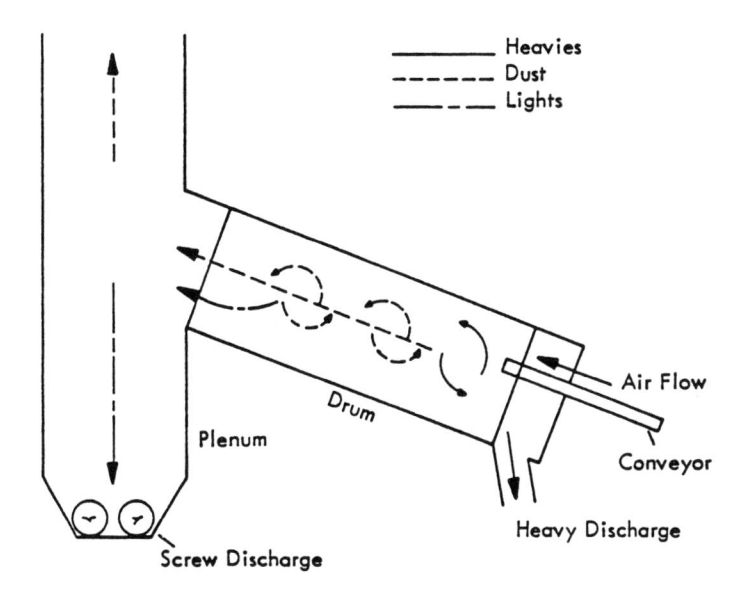

Figure 3-9. Rotary air classifier [57].

acceleration. (Cyclone design is based on the vortex principle.) The solid gas mixture enters tangentially near the top and is forced down an ever-decreasing spiral to the solids outlet trap. The solid material drops out of the gas stream in increasing quantities as the apex of the cone of the cyclone is approached. The gas forms a vortex in the center and travels upward into the air outlet. Usually the finer dust particles are carried out with the gas and a dust filter is required.

Air classifiers rely on the varying physical characteristics of the component items contained in the waste stream to achieve separation. With mixed municipal solid waste, the separation process becomes extremely complex. The auxiliary equipment used with most air classifiers includes a surge bin to level the variable output of a shredder, a metering feed device, a separation chamber, pneumatic transport tubing for the light fraction, a light fraction deentrainment cyclone, an air mover, a final filter to remove the entrained dust from the transport air and a discharge conveyor to transport the heavy fraction.

In addition to air classification systems, a number of other gravity-type separators should be described briefly. Most of these gravity separation systems are wet processers and have been utilized for recovering a variety of fractions in municipal refuse. In most of these processes the solid waste is converted to a liquid slurry. After processing, the recovered fraction is usually dewatered. One hydraulic separation system used for separation

56 DESIGN PRINCIPLES

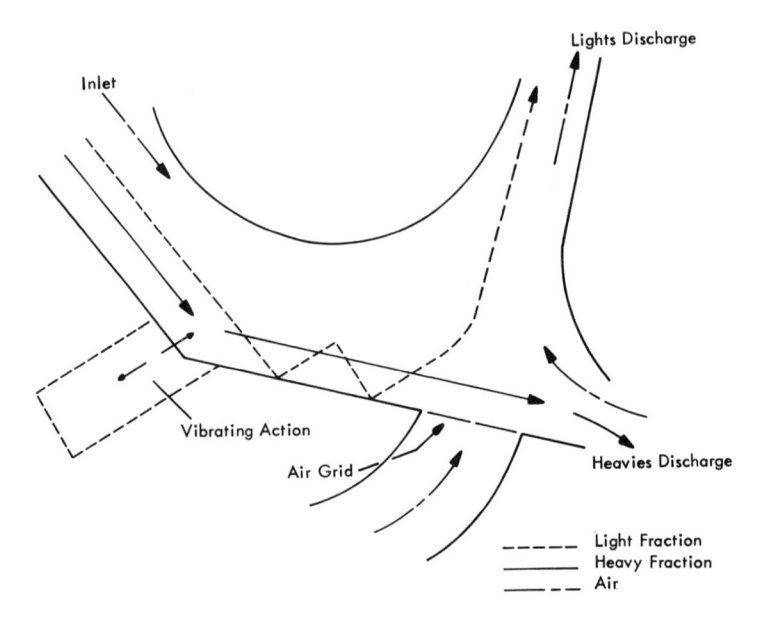

Figure 3-10. Principle of vibrating air classifier [58].

of the inorganics from the heavy organics in the air-classified heavy fraction is the rising current separator. In this unit, the paper, wood and plastics are floated off while the metal, glass and rock sink to the bottom of the tank. A rising current of water is used to provide an effective fluid specific gravity greater than one, thereby floating off those materials having lower sinking velocities. Although density will play an important role, the size, shape and drag coefficient will all affect how a particular component will be classified in this unit. The rising current separator consists of a top section, pump section, feed section, throat or separating section, overflow weir, sink elevator, sink chute and dewatering screen, as shown in Figure 3-11 [59]. The pump circulates water through the upper chamber, where it emerges as a uniform flow and rises through the throat or separating section before spilling over the weir, draining through the dewatering screen back to the pump. Solid waste fed to the separator is submerged in the feed section and transported into the rising current area. Those materials having a velocity less than the effective rising current flow over the weir into the float side of the partition. Material with the greater velocity sinks through the rising current and is picked up by the rotating drum elevator. The drum elevator lifts the sink particles and discharges them down the chute to the screen sink side. The washed or inorganic material recovered in this process then can be transported to a

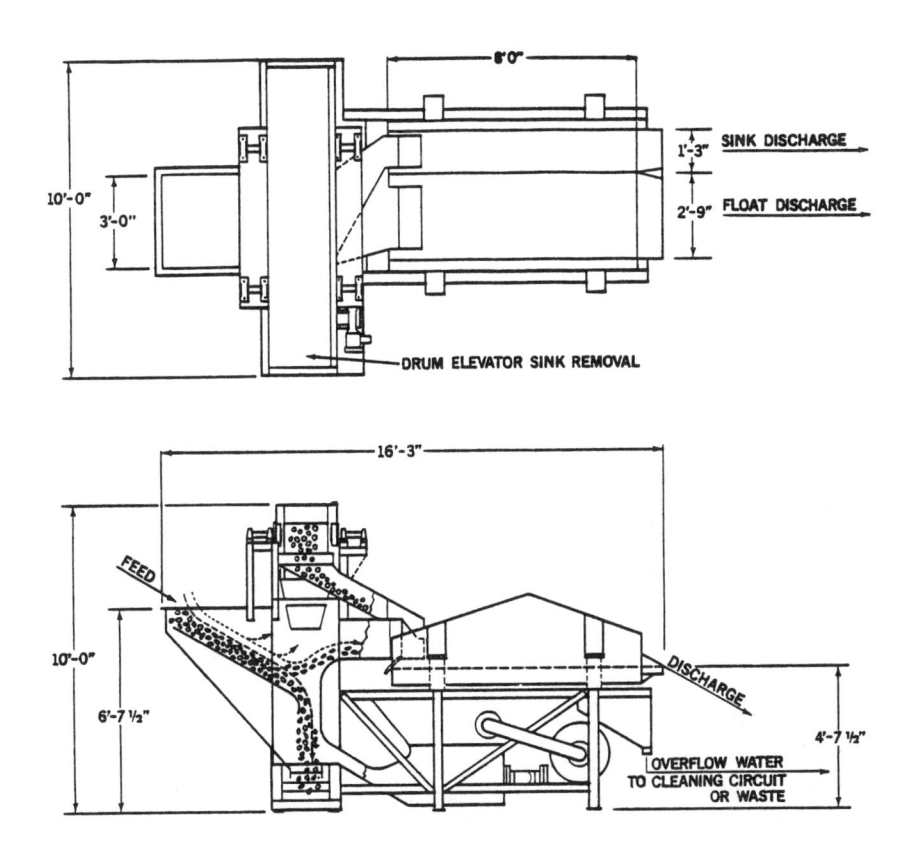

Figure 3-11. Wemco RC separator [59].

second gravity separation facility to recover selected metal and glass fractions [59].

Dense media separation (sink-float process) is an example of a gravity process that could be used to recover the metal and glass fractions. This technique separates material by the use of controlled-density fluids. Materials with a specific gravity higher than that of the medium used will sink to the bottom of the tank, while those with specific gravities less than the medium will float to the top. Dense media separation has been used for a number of minerals benefication applications and can process materials from 3 in. down to 0.066 in. in size. Two types of heavy media are used for these gravity separation processes: (1) organic liquids, and (2) solid-water systems. The solid-water suspension is usually a slurry of finely ground ferrosilicon magnetite or galena powder and water.

The liquid ratio will determine the specific gravity of the medium. Specific gravities of 1.25 to 3.4 have been reported for these materials. For dense media separation to be economically effective the metal powders must be recovered by magnetic separation, and the galena can be recovered by froth flotation. Ferrosilicon is the most widely used metal powder for dense media separation. Organic liquids of selected densities also can be used in this gravity separation process. Chlorinated and brominated hydrocarbons are the compounds used most frequently; however, their use presents a number of drawbacks, particularly those related to health and safety. Dense media separation processes have been used for the recovery of glass and aluminum from the air-classified heavy fraction of municipal refuse [33,60].

Another gravity separation process extensively used in minerals benefication and that is now being applied to solid waste processing is froth flotation. This process is concerned primarily with the separation of finely divided materials. The mineral waste fraction to be processed is first finely ground and then treated with a chemical agent that imparts hydrophobic properties to the fraction of material to be separated out. A water slurry of the treated waste mineral powder is prepared and transported to a processing tank. Air bubbles are introduced into the bottom of the processing tank and, as the bubbles rise through the slurry, the fraction to be separated out is carried to the surface with the bubbles and collected by skimmers. By selecting the proper chemical agent, it is possible to separate the glass fraction from a mixture of waste minerals (stones, pieces of pottery, etc.) [61].

Another means of separating materials by their density is vibration. A mixture of materials subject to vibration forces will tend to segregate by density if the individual particles are essentially the same size. Stoners, jigs and wilfley tables all utilize this principle in separating desired

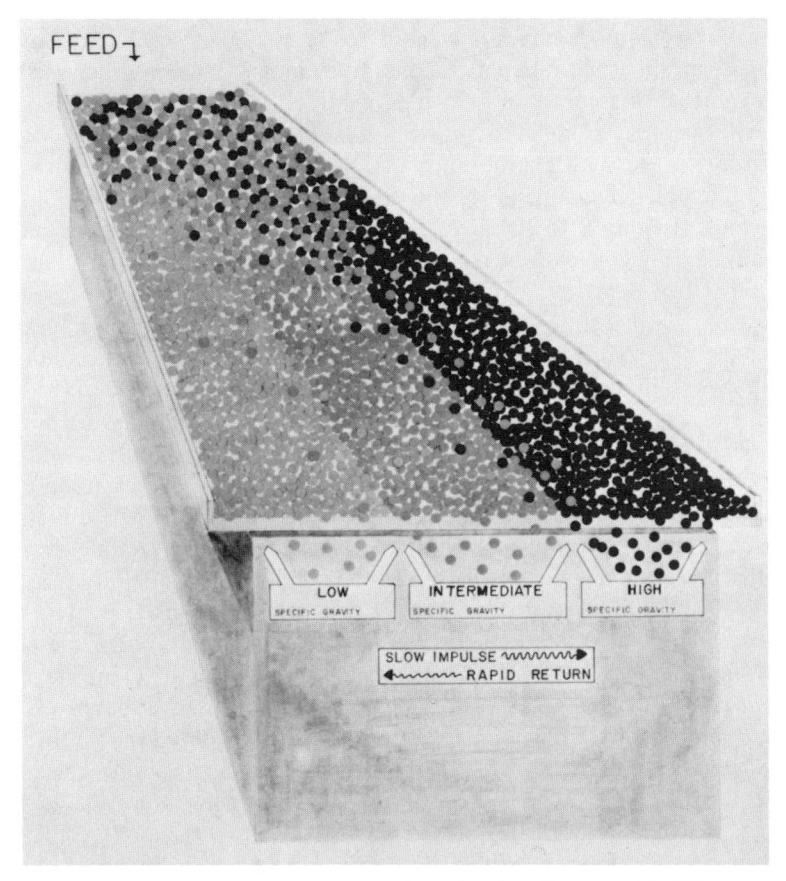

Figure 3-12. Schematic representation of a dry vibrating table [62].

mineral fractions for recovery (Figure 3.12) [62]. In most of these systems, minerals to be processed are vibrated across a table or screen by a system of air jets, water jets or mechanical vibration. By controlling the frequency of vibration and using inclined surfaces, it is possible to separate out the fraction desired for recovery. These systems have been used for separating inorganic materials from the organic fraction as well as separating glass from different metallic fractions in solid waste [33].

3.3.3 Magnetic Separation

Magnetic separation provides another means for recovering products of economic worth from municipal refuse. A variety of equipment is available for separating out the ferrous materials in refuse. Two types of

60 DESIGN PRINCIPLES

magnetic separators commonly used for ferrous recovery are the overhead suspended belt magnet and the drum magnet. When solid waste is transported on a conveyor belt underneath either a drum or belt magnet system, ferrous metals in the waste will be lifted away from the conveyor and carried to a discharge chute for further processing. The magnet manufacturers recommend that when either magnet is used it should be suspended within 8 to 12 in. of the conveyor. They further recommend that the refuse conveyor be traveling at speeds up to 400 ft/min to achieve minimal bed depth (1–3 in.) (Figures 3-13 and 3-14) [63].

In the case of a simple belt-type magnet, the separator usually is located over the discharge end of the solid waste conveyor, and the attraction of the magnet lifts the ferrous metal from the waste and carries it on the belt to the end of the magnetic field, where it is dropped to a receiving bin. A more advanced belt-type magnetic separator has been specially developed for municipal solid waste (Figure 3.14). This suspended self-cleaning-type magnet consists of three separate magnets arranged in a "V" configuration. A rubber belt conveyor moves the lifted material from one magnet to the next. The magnetic fraction in the solid waste is lifted by the first magnet and conveyed around the "V" to the second magnet, which is reversed in polarity to the first. This results in agitation of the material collected and causes trapped nonmagnetics to be released. There is an air gap between the second and third magnets, and the material passing from the second to the third is momentarily released, providing a means for any additional nonmagnetic material trapped with the ferrous to be released. The ferrous material attracted to the third magnet is carried to the discharge end of the belt.

In a drum magnet, a stationary magnet is located inside the revolving drum shell. Ferrous material in the solid waste stream dropping off a conveyor belt is lifted up by the magnet and conveyed around the circumference of the drum until it passes the magnetic field and is discharged. To reduce the amount of nonferrous material collected by a drum magnet, a second drum magnet can be used on the recovered ferrous to obtain a cleaner fraction for further processing. Another option for obtaining a cleaner ferrous fraction is the use of an aspiration system, which will lift away any of the light, paper and plastics that are trapped with the ferrous material.

In most processing systems, the ferrous fraction is usually conveyed to a compactor for densification. On recovery, the ferrous metal usually has a density of 34 lb/ft^3, which can be compacted to about 75 lb/ft^3 for transport to the ferrous processor. Compaction is achieved by both hammermills (minimills) and hydraulic compactors (boilers). Although the amount of ferrous metal in the refuse is variable, there appears to be

BASIC PROCESSING TECHNOLOGIES 61

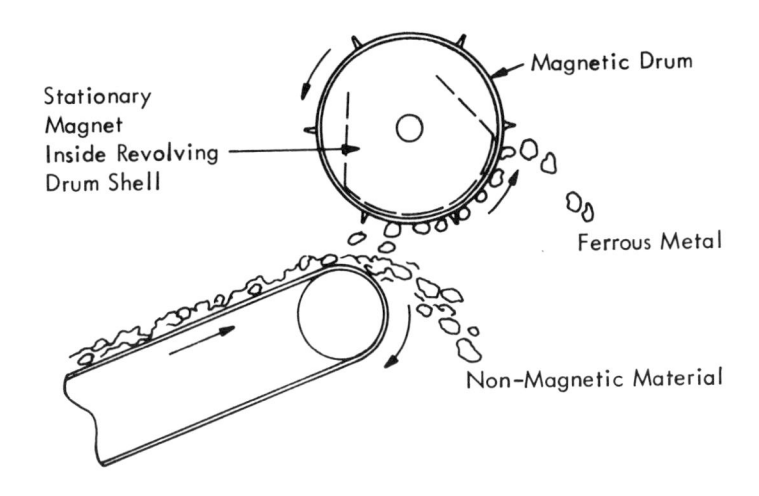

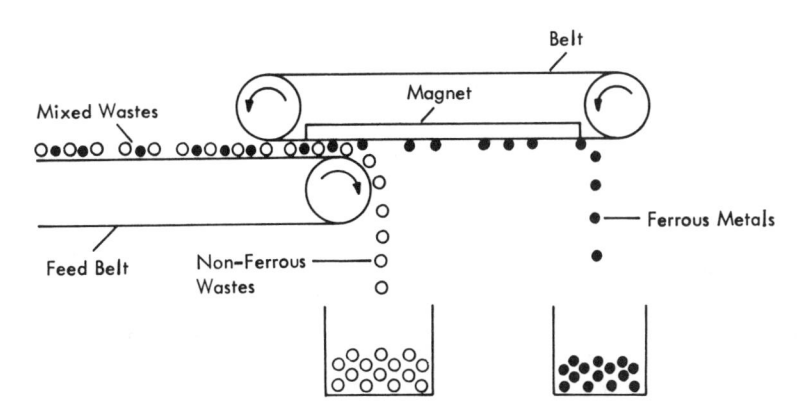

Figure 3-13. Conventional magnetic separation [63].

an average of about 7½% ferrous in municipal solid waste. The available data indicate that about 80–85% of the ferrous metal can be recovered from the municipal refuse by magnetic separation processes. The major components found in the ferrous fraction are tin-coated cans, bimetallic-coated cans and tin-free bimetal cans. These cans represent 60–90% of the ferrous fraction. The remaining ferrous fraction includes wire, bottle and jar caps, the metal ends from paper containers, metal from cookware and hardware, appliances and ferrous metal furniture. Total contaminants found in the recovered ferrous fraction are likely to range from 5 to 15% (trapped food, paper labels, etc.) [19,47,64].

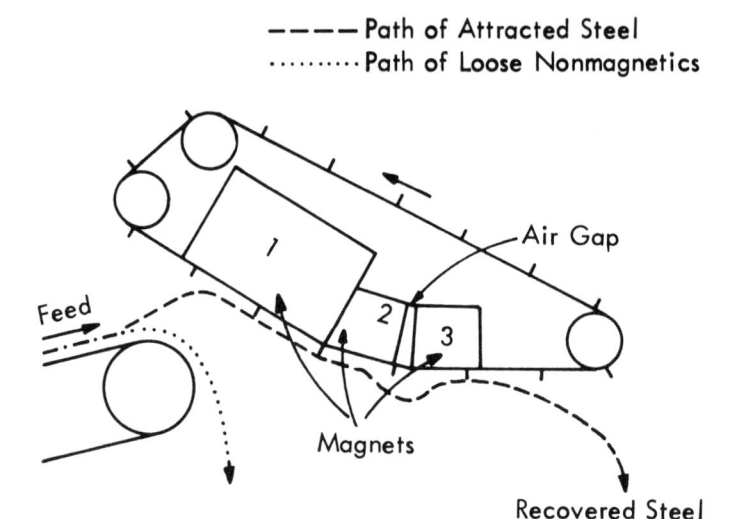

Figure 3-14. Principle of the suspended belt magnet [63].

3.3.4 Other Separation Processes

Although size, density and magnetic separation are the major unit operations used in municipal solid waste processing, there are several other processes that also have been used and are worthy of mention. These processes are designed primarily for the recovery of metal and glass fractions from municipal refuse and depend on the use of electrical or optical techniques. Optical sorting equipment has been developed that can be used to recover clear glass (flint) from an air-classified heavy fraction. The optical sorting devices separate particles of agricultural and food processing applications. A schematic of the sorting process used for glass recovery is shown in Figure 3.15. As shown, the material passes by an optical detector, which activates a compressed air jet to blow aside the unwanted material into a reject chute. This technique can be employed to recover clear glass particles in excess of a ¼ in. in size [65].

Separation of nonferrous metals from the air-classified heavy fraction of municipal refuse is also of considerable interest. Eddy currents and electrostatic separators both offer considerable potential for recovery of aluminum and other nonferrous metals. In the eddy current process, currents are introduced into the nonferrous metals passing on a conveyor belt through the unit. An eddy current flux is produced that results in the generation of a repulsive force, which causes selected nonferrous metal materials to be shunted off the conveyor into a collection bin [66].

BASIC PROCESSING TECHNOLOGIES 63

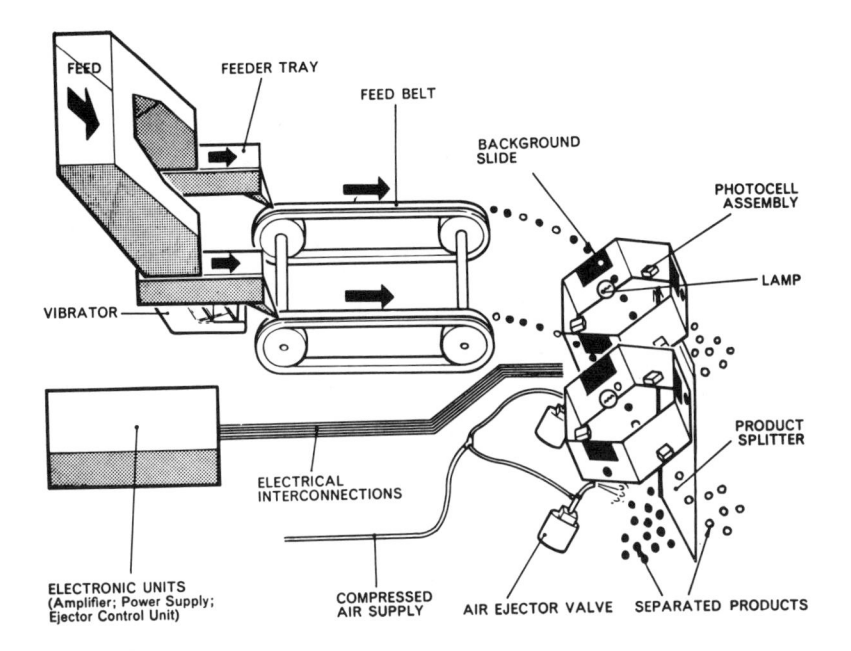

Figure 3-15. Electronic optical sorter [65].

In electrostatic devices, electric charges are applied to nonmetallic materials from a high-voltage discharge, which results in the attraction of these nonmetallic materials to a negatively charged collection surface. The Bureau of Mines utilized this system in the municipal solid waste pilot plant [67].

3.4 MATERIAL HANDLING

In addition to the specific processing operation, the waste material has to be received, discharged, stored and transported between operations. This section discusses material transport, storage, densification and ancillary processes common to almost all unit operations.

3.4.1 Material Transport

A variety of transport processes are available to move material into, away from and between processing stages. Conveyor equipment can be categorized into five different groups (Figure 3-16):

64 DESIGN PRINCIPLES

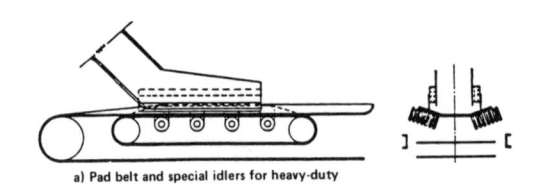

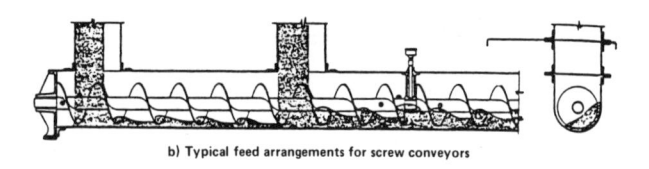

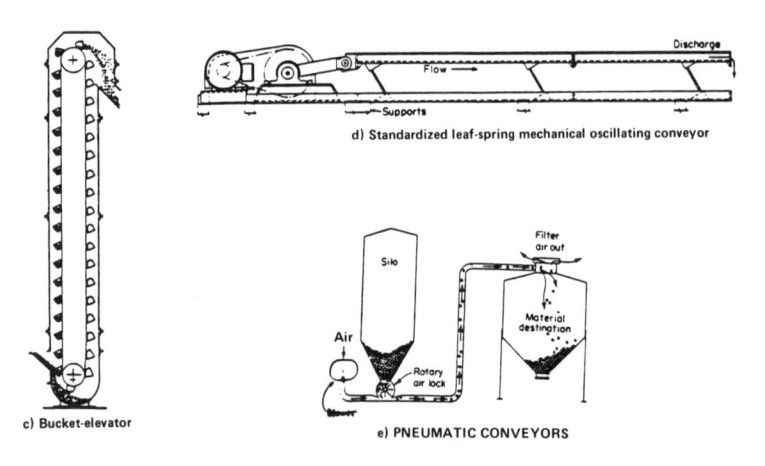

Figure 3-16. Conveyor for bulk solids [68].

1. Belt
2. Flight
3. Screw
4. Vibratory
5. Pneumatic

The equipment selected will depend, to a large degree, on the processing requirements and the transport properties of the material to be conveyed. Density, particle size and shape distribution, abrasiveness, moisture

BASIC PROCESSING TECHNOLOGIES 65

content and flowability (angle of repose) are the major material properties for consideration in conveyor selection. The angle of repose for refuse is in excess of 90 degrees, which may be desirable for stacking but presents many material handling problems. In dealing with refuse, it is important to remember that refuse is a unique composite material containing many different items but it is not characteristic of any of its individual components.

Belt conveyors are belts constructed of several plys of canvas, rubber or nylon around two pulleys. The weight of the material is supported on idlers periodically spaced along the span of the belt. On a flat belt the carrying capacity is limited by the angle of repose of the material to be transported. By inclining the ends of the idlers 20 degrees to create a trough, the capacity of the belt will be increased significantly (Figure 3-17). Belts are the most commonly used conveyors. They can be used to convey materials long distances (miles), up slopes less than 30 degrees (usually 18–20 degrees) at a variety of speeds to 1000 ft/min (normally 60–120 ft/min) and carry loads to 5000 ton/hr. A single rubber belt conveyor can carry material up an incline and on a horizontal plane. However, certain problems are associated with their use. Of most concern are problems caused by material falling from the belt due to back sliding and idle agitaiton. Skirts or dribble pans can be employed to minimize these problems. Belt damage due to sharp or ragged objects inthe waste also can be a problem.

Flight conveyors can be described as metal pans, buckets or flights driven by chain and sprocket. The metal pans, which usually are made of steel, are overlapped to prevent leakage. The shallow metal pan conveyors, termed "apron conveyors," are frequently used as initial feeders in

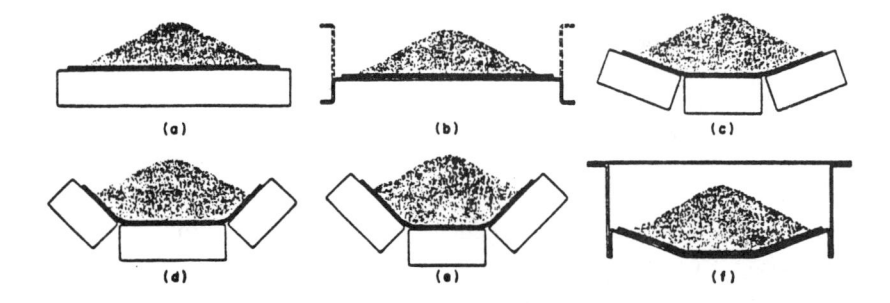

Figure 3-17. Typical belt-conveyor idler and plate-support arrangements: (a) flat belt on flat-belt idlers; (b) flat belt on continuous plate; (c) troughed belt on 20° idlers; (d) troughed belt on 45° idlers with rolls of unequal length; (e) troughed belt on 45° idlers with rolls of equal length; (f) troughed belt on continuous plate (Link-Belt Co.) [68].

solid waste processing facilities. Apron conveyors are ruggedly constructed with good resistance to wear and fire and require little maintenance and can be used to convey material up inclines in excess of 35 degrees. The most common problem encountered with apron conveyors is jamming, which is usually caused by material slippage.

Bucket elevators used for lifting materials vertically also have been used for transporting solid wastes. However, because of the large-sized buckets required for raw refuse, the use of these conveyors has been limited. The bucket elevators are used primarily for shredded refuse and separated refuse components.

Pneumatic conveyors, which transport material by suspending it in air stream, also are used for conveying solid waste. These systems can be operated in either a positive or negative pressure mode. The same principles used for air classification are used for conveying solid waste. In the negative pressure systems, material is pulled through the piping system as dirt is pulled up by a vacuum cleaner. Negative pressure systems are best suited for transporting material from multiple sites to a single site. Positive systems, which push or blow material through, are better suited for transporting material from a single site to multiple sites. Air locks are used to feed and discharge the material transported through the closed system.

Pneumatic systems are being used for transporting both raw and processed refuse. A number of apartment buildings, hospitals and commercial facilities have installed pneumatic systems for transporting raw refuse to a central collection or processing facility. Pneumatic systems also have been used to transport shredded refuse between processing stages and into combustion units. Pneumatic transport provides a closed compact means of moving refuse long distances at relatively high speeds. Elbow errosion, clogging of wet refuse and material leakage are the major problems encountered with pneumatic transport of refuse.

Vibrating conveyors move material by propelling it across a pan. These conveyors operate on the stroke principle, where a force acting at a 20–45 degree angle to the horizontal pan causes material on the pan to jump forward. Vibrating conveyors have been used in a variety of solid waste processing applications, a primary application having been for the uniform feeding of raw and shredded refuse to a processing operation. However, because of the uneven nature of refuse, some problems have been encountered.

Screw conveyors also have been used to a limited extent for refuse transport, mostly for conveying shredded refuse. Screw conveyors help keep the refuse well mixed but are subject to capacity limitation and jamming problems [68,69].

BASIC PROCESSING TECHNOLOGIES 67

3.4.2 Storage

Waste storage is a necessary part of any waste processing system. The processing facility will require an area to receive the refuse and store it until it can be metered into the system. In addition, various stages of the waste processing plan also may require surge storage. Storage is also required for the recovered products, the particular storage system selected being determined by the particular storage requirement and the waste characteristics.

Refuse can be received on an open tipping floor, stored in piles and metered into the system by a front-end loader (Figure 3-18). It also can be received and stored in a pit and metered into the system by an overhead crane and grapple bucket unit. Live bottom pits consisting of a receiving bin with a conveyor or ram unit at the base also are used to receive refuse and continuously meter it into the system (Figure 3-19). During various processing stages, surge bins are required for temporary storage and, after processing, silos are frequently needed to store the refuse-derived products.

Processed refuse can be stored in piles, bins, hoppers and silos. Bins are usually flat or slant-bottomed containers in which the material to be

Figure 3-18. Tipping floor and front-end loader [70].

68 DESIGN PRINCIPLES

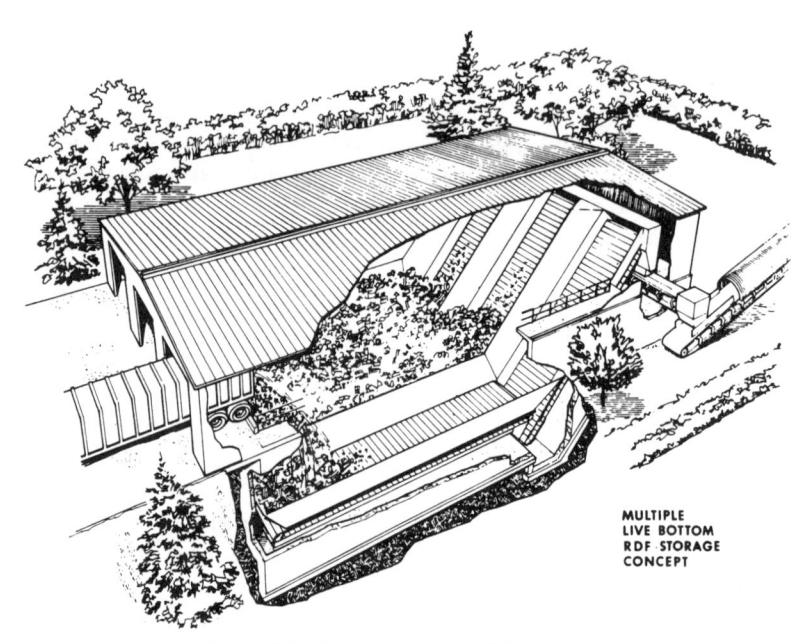

Figure 3-19. Live bottom pit [71].

stored is fed from the top and exited from the base. Hoppers are usually conical bins with sloping bottoms and silos are tall cylindrical bins. A variety of different bin designs are utilized for waste storage, and those selected are frequently those used for wood storage.

Bin design and the condition of the refuse will determine its flow characteristics in the bin. Particle size distribution, density, moisture content and the flow characteristics (angle of repose, the coefficient of friction, etc.) will determine the ease with which the refuse can be stored and recovered.

Three different types of storage bins have been used for the RDF. In the conical "sweep bucket" bin, the RDF is fed in from the top by pneumatic tube or flight conveyor and discharged by a drag chain and sweep buckets. During storage, the refuse forms a large free-standing conical pile in the teepee-shaped bin. The drag chain and buckets circulate around the rim of the waste, with the buckets at the end migrating toward the center, filling and then dragging the RDF to a discharge chute. Two problems encountered with these systems are high wear on the drag chains and bin floor and the densification of the RDF stored in the bin (Figure 3-20).

In the inverted bin, the walls slope inward by four to seven degrees and the bottom is wider than the top. This design tends to reduce bridging and

Figure 3-20. Atlas TP3-type storage (courtesy Atlas Systems).

densification of the RDF. A series of parallel feed screws or rollers arranged at the bottom of the bin are used for material discharge.

A "daffing roll" bin also has been used for processed refuse storage. Screws in the upper portion of the bin serve to level the refuse in the bin. The refuse is discharged by a drag chain system in the floor of the bin. The discharge system fluffs the refuse, giving a more uniform discharge material that is easier to transport [25,68,72].

3.4.3 Densification

Raw refuse can be densified by a variety of hydraulic compaction equipment. Baling units can compact raw refuse to densities of more than 60 lb/ft^3. Shredded and powdered refuse can be densified by a variety of extrusion and compaction processes. In the pelletized form, RDF has several advantages; it can be transported and stored more easily and it is more compatible with stoker-fired furnace units. Processes are available to produce cylindrical and cubical configurations. Extruders used for animal pellet production are the most commonly used equipment for preparing cylindrical pellets. Rectangular and square RDF briquettes are produced using straw cubbeters. In most cases, binding agents are not required; however, a major problem encountered in RDF densification is the high rate of die wear.

3.5 THERMAL PROCESSES

In the treatment of solid waste, a number of processes are used that involve thermal energy. A number of these processes are endothermic and require the input of heat energy, while others are exothermic and give off heat energy. Thermal processes of particular importance in the treatment of municipal solid waste are combustion, pyrolysis and drying.

3.5.1 Combustion

By definition, combustion is the rapid combination of oxygen with a fuel, resulting in the release of heat. Carbon and hydrogen are the two main elements found in all fuels. The oxygen needed for most combustion processes is obtained from air. In the basic combustion process one atom of carbon reacts with two atoms of oxygen to form carbon dioxide, while two atoms of hydrogen combine with one atom of oxygen to form water. The organic fraction of municipal waste in the as-received condition has been defined by Wilson [25] as $CH_{2.45}O_{1.03}$ and would require about four pounds of air per pound of waste for perfect combustion; however, most combustion processes require 100–200% excess air.

A wide range of furnace designs have been developed for the combustion of municipal refuse, either for size reduction (incineration) or energy recovery. Both periodic and continuous furnace units are available for the combustion of raw or processed refuse. Moving grate and rotary kiln-type units are most commonly used for burning raw refuse. Shredded refuse usually is burned in semisuspension-type units (spreader-stockers, etc.). Shredded and powdered refuse also have been burned in conventional coal-fired boilers, comprising from 20 to 50% of the fuel mix (by heat content). Shredded refuse also has been fired successfully in cement kilns, vertical shafts, furnaces, fluidized bed units and high-temperature furnaces (3000°F).

In addition, both raw and shredded refuse also have been burned with sewage sludge (codisposal). Refuse and sludge mixtures have been burned in both refuse and sludge incinerators in the United States and Europe. In the United States, codisposal is still primarily in the developmental stage. The Europeans have had more experience with codisposal programs.

In sludge incinerators the refuse serves as a fuel for burning partially dewatered sludge. In most of these systems, the refuse is processed to RDF and then used in place of oil or pulverized coal in both the multihearth or fluidized bed units. Although this approach has had limited success, the economics appear to favor the combustion of refuse and sludge mixtures in refuse incinerators.

In a refuse combustion system, the partially dried sludge (20-40% solids content) can be injected into the combustion zone (over the grates, etc.) or mixed with refuse prior to combustion. The sludge, which is initially 5-10% solids, is mechanically dewatered and then usually dried using flue gases from the combustion unit. Although the Europeans have had the most experience with codisposal, it is being experimented with in the United States at several facilities. At the Harrisburg incinerator facility, sludge with 25% solids content is mixed with the raw refuse in the pit and fired in the water wall incincerator. The sludge constitutes about 4% of the fuel feed. Dried sludge also has been cocombusted with refuse at the Eastman Kodak Company suspension-fired industrial boiler. Although there are a number of benefits with cocombustion of refuse and sludge, there are still a number of technical problems to be overcome before the concept will be economically effective. In addition to the expected processing-type problems (clinker formation, etc.), there are also environmental concerns about the by-product from the cocombustion of refuse and sewage sludge (heavy metals, etc.) [73-78].

3.5.2 Pyrolysis

Pyrolysis is defined as the thermal decomposition of a substance in the absence or partial absence of oxygen. Pyrolysis of refuse results in the production of bituminous-like liquid, a char residue and gaseous product consisting primarily of carbon monoxide, hydrogen, methane and carbon dioxide, and water vapor.

An effective insight into the pyrolysis of municipal solid waste can be obtained by studying the thermal decomposition of cellulose and cellulose products. The pyrolysis of cellulose can be viewed as a two-stage process: (1) dehydration and (2) depolymerization, although there does not appear to be a sharp delineation between the two. Below 390°F, the predominant reaction is dehydration. Water vapor and traces of carbon dioxide, formic and acetic acids, and glyoxal are evolved. Between 390°F and 535°F, both dehydration and depolymerization reactions occur. Larger quantities of carbon dioxide, formic and acetic acids, and glyoxal are evolved at these temperatures. In addition, small amounts of carbon monoxide may be released. Above 535°F, the primary pyrolysis reaction is depolymerization, achieved by scissions of C-O bonds in the cellulose chain, either in the rings or between them. Scissions of C-O in the rings result in the disintegration of the ring to yield CO_2, CO and H_2O. Scissions of the C-O bonds between rings results in the production of levoglucosan molecules. Depending on the reaction conditions, levoglucosan either may volatilize or decompose thermally to yield gases and a carbonaceous material. The main volatile products from levoglucosan decomposition are CO, H_2, CH_4, CO_2, acetic acid, ethanol, acetaldehyde, acetone, biacetyl, methylethyl ketone, ethylacetate and tars. Above 930°F, secondary decomposition and gasification of the char occur. A summary of the pyrolysis reactions for cellulosic materials is presented in Table 3-3.

As indicated, organic liquids, char, water and a gas are produced from the pyrolysis of cellulose. The quantities generated are controlled by the heating rate, final temperature and length of exposure to final temperature. In general, the char will constitute between 20 and 40% of the final product mix, the organic liquids and gas phase can vary between 10 and 40% and water constitutes the remaining fraction. The thermal process employed can be designed to maximize the end products desired by the cellulose wastes. Higher heating rates and temperatures produce larger quantities of gas and less char. Conversely, lower heating rates and temperature processes result in increased char production. The quantity of char, the composition of the gases and liquids evolved and the necessary reaction temperatures can be affected by the presence of chemical agents in the cellulose materials. A number of chemical compositions have been identified that can control the quantity of char, combustible gases and tars produced [79].

Three basic designs have been used for the pyrolysis of municipal refuse: horizontal shafts, rotary kilns (inclined horizontal shaft) and fluidized beds. The shaft-type reactor units are the least expensive and the simplest to operate. The refuse is introduced and, by rotational gravity, progresses to the lower end, where it is discharged. The gases and liquid vapors generated defuse up through the reactor and discharge through the

Table 3-3. Pyrolysis Reactions of Cellulose

Temperature (°F)	Process	Major Volatile Products
<390	Dehydration	Water vapor
390–535	Endothermic dry pyrolysis	CO_2, water vapor and acetic acid
535–930	Exothermic pyrolysis to char	CO, H_2, CH_4, CO_2, acetic acid, ethanol, acetaldehyde, acetone, biacetyl, methylethyl ketone and tars
>930	Gasification of char	HCHO, H_2 and CO

top. Both direct and indirect heating methods are employed for the pyrolysis reaction. To date, neither the shaft-type units nor the fluidized bed reactors have been effectively commercialized.

3.5.3 Refuse Drying Processes

As shown, the moisture content of municipal refuse will vary from 15 to 20 wt % of the refuse and averages about 30%. This high moisture content inhibits refuse separation processes and reduces Btu content. Furthermore the large variation in moisture content from day to day adversely affects the commercial marketability of the refuse-derived fuel. It is desirable, therefore, to consider drying the refuse as part of the resource recovery processing system. A number of different processes can be considered for reducing and controlling the moisture content of raw and processed refuse.

Freeze-drying, vacuum drying, chemical drying, and thermal drying and mechnical dewatering processes can be used to reduce moisture content. Both freeze-drying and vacuum drying are prohibitively expensive. Several different chemical drying techniques developed in the laboratory also have been considered for moisture reduction but currently are not commercially available. Based on the experience to date, it would appear that, of the various processes considered, thermal drying is the most practical means for removing moisture from refuse. At present, three different types of thermal dryer designs are available for processing municipal solid waste: (1) rotary drum, (2) fluidized bed, and (3) revolving tray. Based on limited cost and operational data available for these units, it is not possible to compare these different designs for the processing of municipal solid waste.

Both direct and indirect heating processes can be utilized for moisture removal in these dryer designs. Indirect heat dryers heat the waste by radiation, while direct heat dryers bring the waste into direct contact with the drying medium, which can be hot air, steam or hot exhaust gases from a combustion process. All three heating systems have been employed for the drying of municipal solid waste. Because of the limited use of refuse drying systems, a number of questions remain concerning their performance. In addition, the costs and benefits of utilizing a dryer in the processing of solid municipal waste have yet to be clearly established.

Moisture also can be removed from MSW by mechanical processing. However, most of these processes (filter press, hydrocone, centrifuge, etc.) are designed primarily for reduction of water from slurries where the solids content is only 5 or 10%. In addition, water removal to levels below 40–50 % by mechanical processes is very costly.

3.6 CHEMICAL AND BIOLOGICAL PROCESSES

A variety of both chemical and biological processes are used in the treatment of municipal solid waste. Most of these processes are used for converting the organic fraction of waste into useful products. In addition, a number of chemical processes are used in combination with thermal treatments to enhance the quality of products derived from the organic fraction in municipal solid waste. The chemical treatment of primary interest is acid hydrolysis for the conversion of cellulose waste to glucose. In addition, chemical treatment for cellulose embrittlement, enhanced carbonization, hydrogenation and partial oxidation processes also are being developed for municipal solid waste treatment. Anaerobic digestion for methane production, enzymatic hydrolysis for glucose production, aerobic digestion for humus and fermentation for alcohol and protein production are the major biological processes of interest.

3.6.1 Acid Hydrolysis

The cellulose waste fraction in municipal refuse can be hydrolized in the presence of an acid to form glucose:

$$C_6H_{10}O_5 + H_2O \rightarrow C_6H_{12}O_6$$

Two basic processes for the acid hydrolysis of cellulose are used for the production of sugar: (1) dissolving cellulose in concentrated acid followed by dilution and distillation, and (2) hydrolysis in dilute acid solution at elevated temperatures and pressures followed by separation of the acid and sugar. In these processes the acid serves as a catalyst and the glucose can be fermented to form alcohol. Acid hydrolysis of the cellulose fraction in municipal refuse has been studied in the laboratory and demonstrated on a pilot–scale. Both the laboratory-scale studies and pilot-scale experiments were based on a plug flow reaction model. A dilute mineral acid solution at elevated temperatures and pressures was employed. The organic fraction of the refuse was slurried and treated with a 0.4–0.6% mineral acid at 450–500°F using a residence time of 5–30 seconds. A glucose yield of 55% was obtained at temperatures of 450°F using a 0.4% acid treatment. With the increased interest in cellulose materials for alcohol production, greater support is quite likely for the conversion of the cellulose fraction in refuse to glucose for fermentation to alcohol [76].

3.6.2 Cellulose Embrittlement and Enhanced Carbonization

The organic fraction in municipal refuse can be converted to a powder by a number of chemical/thermal treatments. Both mineral acids and oxidizing agents can be used to embrittle cellulose materials. Chlorine, nitric acid, sulfuric acid and hydrochloric acid treatment in combination with low levels of heat have been found to be very effective for converting shredded cellulose waste to a friable material, which can be easily powdered [46].

A number of chemical treatments for enhancing the carbonization of municipal solid waste to a powdered char also have been identified. These chemical treatments increase the quantity of char produced, decrease the amount of combustible gases and tars formed, and reduce the reaction temperature required for the process. Salts of strong acids or bases and several oxidizing agents have been found to be effective for enhancing carbonization. Recent studies have shown that a 2% solution of sodium aluminate promotes carbon formation of shredded municipal refuse at temperatures below 660°F. Refuse treated with a 2% solution of sodium aluminate would then be carbonized in a rotary kiln reactor unit. A plant processing a 1000 ton/day of raw refuse could generate approximately 300 tons of good quality char. To date, the study of carbonization of municipal refuse has been at bench scale only [80].

3.6.3 Hydrogenation

In the hydrogenation process, hydrogen is reacted with the organic fraction of refuse under elevated temperatures and pressures to produce synthetic gaseous liquid fuels. A number of hydrogenation processes have been developed and demonstrated on a pilot scale using the organic fraction of municipal solid waste as the feedstock. As described in Chapter 1, the Bureau of Mines and the Worcester Polytechnic Institute hydrogenation processes were developed for the conversion of municipal solid waste (see Section 1.7) [81].

3.6.4 Biological Conversion Processes

A number of biological conversion processes have been developed for converting the organic fraction or municipal refuse into useful products. Compost, methane, glucose, alcohol and protein have been obtained from the organic fraction of municipal refuse by the use of biological agents.

Aerobic digestion is one of the most common biological treatments used in the processing of refuse. In this process, the organic fraction of the refuse is converted to a compost by microbial digestion at temperatures of 150–170°F. Both mechanical digestion tanks and open field windrows (100-foot-long piles, 8 feet wide, 5 feet high) are used for composting treatments. The mechanical digestion tanks require about 5 days treatment, while the windrows require about 30 days of treatment time. The material in the windrow must be agitated frequently to allow adequate air contact during curing. The microbes for the digestion process are provided by additions of sewage sludge to the refuse. Usually the refuse is shredded and the nondigestible material removed prior to being mixed with the sludge for curing. Unfortunately, the composting process generally has not proved to be economically effective [72,78].

Anaerobic digestion, unlike aerobic digestion, occurs only in the absence of oxygen. In this biological process, the shredded organic fractions of refuse are mixed with sewage sludge containing bacteria, which converts the organic materials to gaseous products. The basic reactions result in the production of methane. The temperature, pH, atmosphere and composition of the refuse will all effect the parameters of the digestion process. Both low-temperature (mesophilic 95–115°F) and high-temperature (thermophilic 130–140°F) range processes are known. In the thermophilic range, lower process times, increased reaction rates and smaller volumes are required.

Slurries with a pH of 6.6–7.6 are used, but best results are obtained with values of 7.0–7.2. A nonoxygen atmosphere is required, and aerobic conditions will result in process instability. It is also important to note that the composition of the feedstock will greatly affect the digestion process. The different types of materials present (proteins, carbohydrates, lipids, salts, etc.) serve as substrates for the different bacterial agents in the sewage sludge. However, toxic substances in the waste (sulfides, heavy metals, etc.) can poison the microbes and destroy the digestion process.

The processing time required for the process can vary considerably. Depending on the specific conditions, from 4 to 20 days may be required for the complete process. Both single- and double-stage digestion processes are employed. The gas generated during a steady-state digestion process must be removed continuously from the digestion tank. The gas collected is usually processed to remove the CO_2, H_2S and H_2O prior to compression for injection into the pipeline [76].

Fermentation of the cellulose waste by enzyme systems produced by the *trichoderma viride* are effective for the hydrolysis of cellulose to glucose. Although the glucose produced is relatively pure, the process is very slow and subject to contaminants that can poison the bacteria. The effectiveness of conversion will vary and the percentage of saccharification can range from 10 to 50%, depending on the nature of the cellulose waste. The most extensive work to date for enzymatic hydrolysis of municipal solid waste has been by L. A. Spano of the United States Army Natick Laboratory. A schematic of the basic process used is presented in Figure 3-21 [82], which shows that the glucose syrup obtained can be further fermented to single-cell protein, alcohol and other fermentable products (butanal, acetone, etc.) [81].

The glucose hydrolized from the refuse can be fermented by properly selected yeast to alcohol. In this process, a 15% glucose solution with yeast is pumped into fermentation tanks. The basic fermentation process, which can take up to four days to go to completion, follows this reaction:

$$C_6H_{12}O_6 \rightarrow 2C_2H_5O_8 = 2CO_2.$$

The pH of the glucose solution is maintained between 4 and 5, and the initial processing temperatures are maintained at 70–80° F; 45–55% of the glucose solution is fermented to ethanol, which is then distilled for use. The residues from this process have been used as animal feed and fertilizer. To date, conversion of municipal solid waste to alcohol has only been demonstrated in the laboratory and at pilot scale [81].

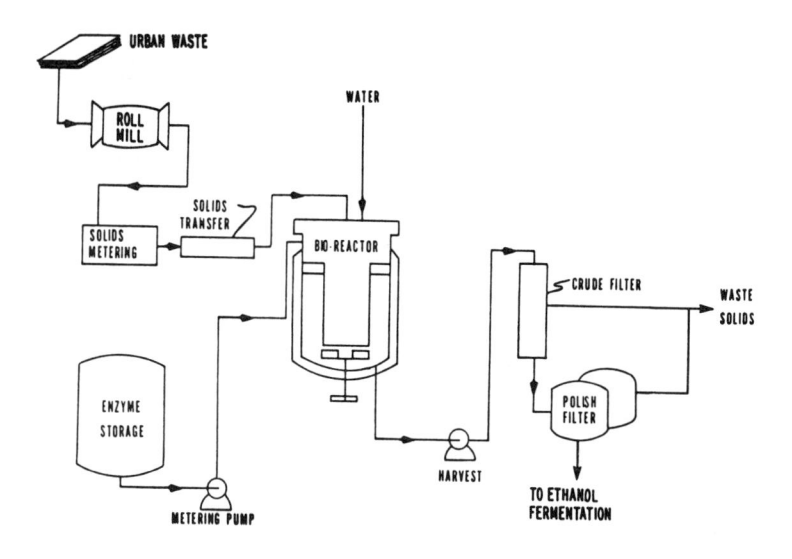

Figure 3-21. Process design of pilot plant for hydrolysis of urban waste, based on U.S. Army Natick Research and Development Command, Natick, Massachusetts [82].

SUMMARY

In this chapter diverse unit operations for processing municipal solid waste have been described and processes for size reduction, component separation, conversion of the organic fraction and material handling have been discussed. These unit processes can be coupled in a variety of different combinations for a range of resource recovery systems; however, the design, performance information and level of understanding available for each specific unit operation are quite variable. In some cases extensive data are available; however, in others, the data are quite fragmentary. Overall, the state-of-the-art is in need of considerable development.

CHAPTER 4

PLANT DESIGN

4.1. INTRODUCTION

Plant design can be properly defined as those processes that culminate in the development of a floor plan organizing the required unit operations into an efficient arrangement and the development of a building design necessary to effectively house the unit operations and auxiliary systems. Although resource recovery facilities are somewhat unique, most of the general procedures developed for plant design can be applied to these systems. It is the goal of this chapter to describe the basic plant design procedures and describe how they are used in the design of resource recovery facilities.

In *Plant Layout and Material Handling* [83], Apple has enumerated eight major objectives for the plant design process:

1. Facilitate the manufacturing process.
2. Minimize material handling.
3. Maintain flexibility of arrangements and of operation.
4. Maintain high turnover of work in process.
5. Hold down investment in equipment.
6. Make economical use of the building cube.
7. Promote effective utilization of manpower.
8. Provide for employee convenience, safety and comfort in doing the work.

These objectives culminate in the goal of choosing an effective material flow pattern to economically produce the required commodities. Apple also has enumerated a sequence of 20 steps that can be followed for effective plant design, regardless of the type of facility being planned. These procedures are enumerated and discussed in Section 4.2. The application of these procedures to the design of resource recovery plants is discussed in the subsequent sections of this chapter.

4.2. PLANT DESIGN PROCEDURES

The goal of any plant design process is to facilitate a manufacturing process in the most effective manner possible. This can be accomplished by the development of a floor plan that arranges the physical facility in a configuration that maximizes operating efficiency, permits the efficient flow of materials, minimizes material handling and provides for the most economical use of building space.

The design of any facility can best be accomplished by following an orderly set of procedures. The twenty procedures developed by Apple are enumerated in Table 4-1 and discussed below.

4.2.1 Procure Basic Data

The first step in any design process or, for that matter, any type of planning project, is the acquisition of the required information to develop an effective plan. Each different type of manufacturing process will require different information specifics; however, certain categories of information will be common to all plant design projects. A listing of the more common categories of basic data required for all plant design projects is presented in Table 4-2.

It is important that accurate information be acquired for these items, as well as for the particular manufacturing operations. Without these basic data it is not possible to establish an effective plant design.

4.2.2 Analyze Data

Once the data are assembled, they have to be evaluated to identify and properly characterize the starting materials and the processes required,

Table 4-1. Standard Sequence of Design Procedures

1. Procure basic data.	10. Design activity interrelationships.
2. Analyze basic data.	11. Determine storage requirements.
3. Design production process.	12. Plan service and auxiliary activities.
4. Plan material flow pattern.	13. Determine space requirements.
5. Consider general material handling plan.	14. Allocate activities to total space.
	15. Consider building types.
6. Calculate equipment requirements.	16. Construct master layout.
7. Plan individual work stations.	17. Evaluate, adjust and chenck layout with appropriate personnel.
8. Select specific material handling equipment.	18. Obtain approvals.
9. Coordinate groups of related operations.	19. Install layout.
	20. Follow up on implementation of layout.

consistent with the goals and objectives of the plant design project. In the data analysis step, the basic information needed for the remaining eighteen steps is organized and categorized. The accuracy and completeness is determined and, where necessary, further data are acquired. This bank of information is then reviewed to assess the current "state-of-the-art." It is this analysis that provides the basis for identifying the processes to be employed, the available equipment, the building designs possible, and the health, safety and environmental requirements needed to comply with current and anticipated codes.

4.2.3 Design Production Process

In this step, the specific operations to be performed are selected, and the sequence of processing steps to be utilized are determined. With the starting materials and final product requirements known, it is possible to identify all the processing operations required to convert the material into the finished product. The objective of this phase is to identify the most effective operations to be used and the most cost-effective sequence in which they can be employed.

4.2.4 Plan the Material Flow Pattern

Once the processing steps are known, it is possible to develop the material flow pattern. The flow plan graphically defines the path that the material would take from receiving to shipping the product to the customer. An example of a typical flow plan for a resource recovery operation is shown in Figure 4-1. This flow plan establishes the layout to be used for the proposed plant. The flow plan also shows the major and minor processing operations the material will receive and traces the steps it will follow in going from operation to operation. The flow plan is a major basic step in developing an effective processing procedure and plant layout and should allow for flexibility in the processing system to

Table 4-2. Basic Data Required for Plant Design

1. Composition and characterization of starting materials and components.
2. Product requirements and specifications
3. Required processing operations
4. Market forecast
5. Available processing system/flow plans and equipment
6. Zoning regulations and building codes
7. Applicable Occupational Health and Safety Act (OHSA) requirements
8. EPA requirements

take advantage of changes in the market or potential advancements in the technology.

4.2.5 Consider General Material Handling Plan

The material handling plan converts the static flow plan into a dynamic plan. It defines the procedures and the means by which the material will be conveyed through each processing operation and from operation to operation. The plan defines both the manual and mechanical methods that will be used to take the material from receiving to shipping.

4.2.6 Calculate Equipment Requirements

Once the flow plan is established and the processing operation determined, it is then necessary to identify the equipment requirements. The first step in determining equipment requirements is the development of a material balance for each processing stage. This is done by first characterizing the starting materials and then tracing the changes to which the materials will be subjected as a result of the different processing operations. Frequently, both material composition and quantity will be altered by the different processing operations. In addition, materials/components are joined or residues are rejected. The changes experienced by the starting material as it goes from processing stage to processing stage are a function of the basic material characteristics, the nature of the processing operations and the processing rate.

The chemical composition and density of the starting materials are known at the initiation of the process. The processing rate is determined as a function of the quantity to be processed (production rate), the number of processing lines required (a balance between economy of scale and design flexibility) and the processing capabilities of available equipment. The development of the material balance then becomes a trial and error process, with the material quantity and composition changing as different plant processing rates, number of processing lines and different equipment capacities are tried. The material balance developed for the flow plan shown in Figure 4-1 is presented in Table 4-3.

The trial and error process used for establishing the material balance provides the initial basis for identifying the equipment needed in each processing stage. The material balance, coupled with a knowledge of equipment size and performance and information about production rates required, facilitate the development of preliminary equipment specification. All the items of equipment required for the plant can be determined from the flow plan.

PLANT DESIGN 83

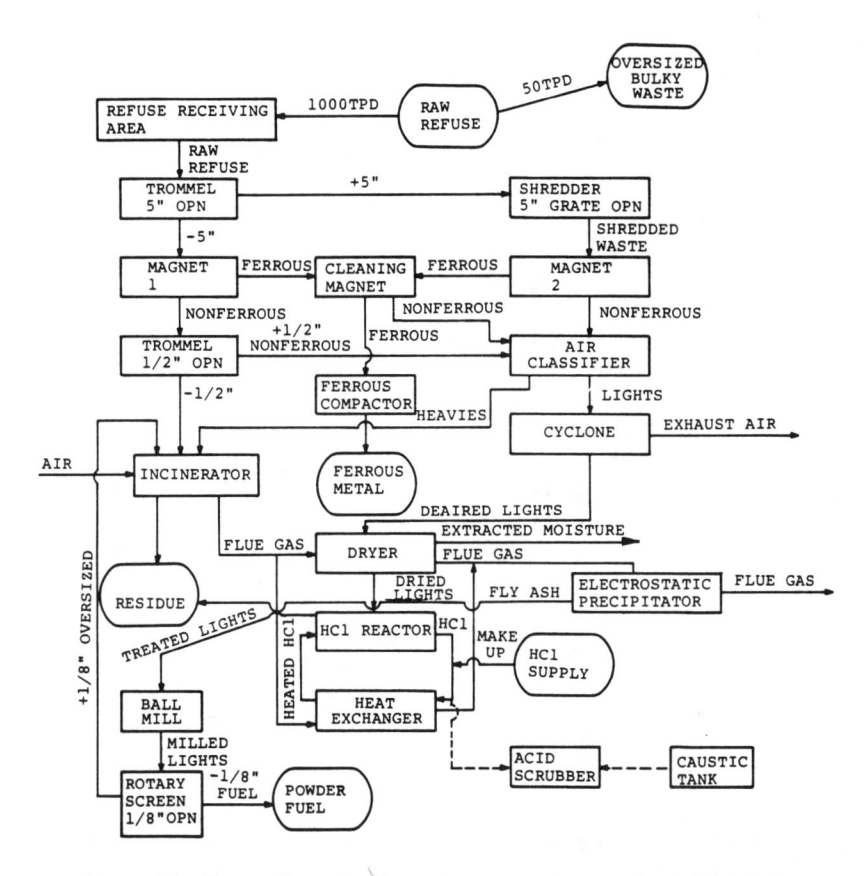

Figure 4-1. Process flow plan for a plant to produce powdered RDF [46].

In the final step to develop equipment requirements, a master chart is prepared itemizing each piece of equipment and the capacity required in each unit operation. This final step may best be carried out with the assistance of equipment brochures and in working with representatives from the various equipment companies. The equipment requirements established also will provide the basis for determining space and labor requirements.

4.2.7 Plan Individual Work Areas

Each individual operation or unit process can be defined as a work area, and in this step a detailed plan is developed for the individual work areas. In addition, the interrelationship of equipment/operators/auxiliary equipment is determined. In the plan developed for each work area,

84　DESIGN PRINCIPLES

Table 4-3. Material Balance [46]

Process	Paper and Paper Products	Plastics	Textiles	Wood	Food and Garden Wastes	Glass	Sand and Rock	Ferrous Metal	Non-ferrous Metal	Total	Destination	Process notes
Trommel												
(5-in. opening)	500	45	15	60	180	70	45	75	10	1000	Entering	(4500 Btu/lb)
=5-in.	330	22	5	15	50	5	5	20	2	454	To shredder	
-5-in.	170	23	10	45	130	65	40	55	8	546	To magnet[a]	
Shredder	330	22	5	15	50	5	5	20	2	454	To magnet[b]	
Magnet	170	23	10	45	130	65	40	55	8	546	Entering	
Ferrous	2	1	1	—	4	—	—	45	2	55	To cleaning magnet	
Nonferrous	168	22	9	45	126	65	40	10	6	491	To ½-in. trommel	
Magnet	330	22	5	15	50	5	5	20	2	454	Entering	
Ferrous	4	1	1	—	2	—	—	17	0.5	22.5	To cleaning magnet	
Nonferrous	326	21	4	15	48	5	5	3	1.5	428.5	To air classifier	
Cleaning magnet	6	2	2	—	6	—	—	62	2.5	80.5	Entering	
Ferrous	1	—	2	—	2	—	—	62	1.5	66.5	To ferrous compactor then to ferrous bin	
Nonferrous	5	2	—	—	4	—	—	—	1	14	To air classifier	
Trommel												
(½-in. opening)	168	22	9	45	126	65	40	10	6	491	Entering	
=½-in.	168	21	8.9	43	69	65.5	13	9.9	5.8	403.9	To air classifier	
-½-in.	—	0.2	0.1	2	57	0.5	27	0.1	0.2	87.1	To incinerator	
Air classifier	499	44.8	14.9	58	121	69.5	18	12.9	8.3	846.4	Entering	
Heavies	38.6	4.3	2.4	28	25	69.0	12.5	10.5	6.6	196.9	To incinerator	
Lights	460.4	40.5	12.5	30	96	0.5	5.5	2.4	1.7	649.5	To dryer	
Dryer	460.4	40.5	12.5	30	96	0.5	5.5	2.4	1.7	649.5	Entering	Ash, 6%; total moisture, 24%; moisture evaporated, 19.6%; heat for drying, 367.7 × 10⁶-Btu
Moisture	65	5	2	4	50	—	1.0	0.5	—	127.5	Exhausted	
Solids	395.4	35.5	10.5	26	46	0.5	4.5	1.9	1.7	522	To reactor	
Reactor	395.4	35.5	10.5	26	46	0.5	4.5	1.9	1.7	522	Entering	
HCl pickup	403.5	36	11.75	27.8	47	0.5	4.5	1.9	1.7	534.6	To ball mill	(HCl gain 12.6)

PLANT DESIGN 85

Ball Mill	403.5	36	11.75	27.8	47	0.5	4.5	1.9	1.7	534.6	To rotary screen
Rotary Screen	403.5	36	11.75	27.8	47	0.5	4.5	1.9	1.7	534.6	Entering
=⅛-in.	40.3	18	2.35	7.0	22	0.25	2.5	1	1	1	94.4
-⅛-in.	363.2	18	9.40	20.8	25	0.25	2.0	0.9	0.7	440.25	To fuel storage
Incinerator	78.9	22.5	4.8	37.0	104	69.8	42	11.6	7.8	378.4	Entering
											To incinerator
Moisture	8	1.5	1	5	52	2	1	0.3	0.2	71	(−6,516 × 10⁶ Btu's)
Ash	2.6	2.3	0.1	1	1.3	67.8	41	11.3	7.6	135	Air required at 100%
Organics	68.3	18.7	3.7	31.0	50.7	—	—	—	—	172.4	excess = 37 8.35 × 2 × 3
Fly Ash	1.3	—	0.3	1.0	1.0	4	3	0.5	0.5	11.6	= 2270 tons
											(18.8%)
Bottom Ash	3.6	2.3	0.6	1.5	3.3	63.8	38	10.8	7.1	131.0	(35.6%)
											(45.6%)
Combustibles, organics											(3%) to electrostatic precipitator (95% trapped)
and HCl (2.2)	66	18.7	2.95	29.5	47.7	—	—	—	—	164.8	(34.6%) to landfill with
Btu's × 10⁶	1020	430	58	475	590	12	7	17	11	2620	95% fly ash-141.95
											(43.6%) total gas
											2521 ton/day
											(3460 Btu/%) about 60%
											available for drying at
											1560 × 10⁶ Btu

the operator cycle, machine operations, economy of motion and material handling requirements are established. In the final phase of this step, the material flow plan through the work area is determined and integrated into the overall flow plan to establish the preliminary space layouts.

4.2.8 Select Specific Material Handling Equipment

In this step of the design process the specific material handling equipment is identified based on the equipment specifications established for each work area. It is important that the material handling equipment be compatible with the processing equipment selected and facilitate effective transfer of the material among the individual work areas.

4.2.9 Coordinate Groups of Related Operations

The process of coordinating groups of related operations is actually initiated in steps four and seven, when the flow pattern is established and the individual work areas are planned. However, it frequently is not completed, and time must be allocated at this point in the project to effectively coordinate the operations of related work areas. This task can be facilitated by the use of a layout planning chart (Figure 4-2), which identifies the major steps required for each operation and serves as a guide to correlate and coordinate interrelated processes.

4.2.10 Design Activity Relationships

Associated with the operations of every plant are a variety of activities necessary for, or ancillary to, the production functions. These activities can be divided into four categories:

    A. Administration
    B. Production
    C. Personnel
    D. Physical plant

The first task is to identify the activities associated with each category, which can be achieved with the aid of a chart itemizing all proposed and ancillary processes and the activities needed to carry them out effectively (Table 4-4). For example, under administration are found management, accounting, engineering, data processing and sales functions. Under personnel there could be health and medical services, food services, lavatories, parking, etc., and under physical plant services there are HVAC, power, water telephone utilities, maintenance, fire protection

Sheet_____ of _____

Part No.__________ Part Name__________ Pcs./Assy.__________ Pcs./Hr. Req.__________

Assy. No.__________ Assy. Name__________ Assy./Job__________ Prod. Hrs./Day__________ Prepared By__________

Material__________ Size__________ Pcs./Day__________ Model__________ Date__________

ST. NO.	DESCRIPTION	OPER. NO.	DEPT. NO.	T.S. REG. NO.	TIME PER PC.	PCS. PER HOUR	OPER. PER MACH.	TOTAL MAN-POWER	MACHINE OR EQUIPMENT	NO. MACH. REQ'D.	DIST. MOVED	HOW MOVED	TYPE OF CONT'R.	REMARKS

Figure 4-2. Layout planning chart [83].

88 DESIGN PRINCIPLES

Table 4-4. Plant Service Activities [83]

A. Administration	a. General	9. Fire escapes
1. President	b. Maintenance	10. Drinking fountains
2. General manager	5. Receiving	11. Telephones—booths,
3. Sales and advertising	6. Stock room (storage)	etc.
4. Accounting	7. Warehousing	
5. Product engineering	8. Shipping	D. Physical Plant
6. Purchasing	9. Tool room	1. Heating facilities
7. Personnel	10. Tool crib	2. Ventilating equipment
8. Product service	11. MH equipment storage	3. Air conditioning equip-
9. File room	12. Supervision	ment
10. Conference room		4. Power generating
11. Vault	C. Personnel	equipment
12. Reception room	1. Health and medical	5. Telephone equipment
13. Switchboard	facilities	6. Maintenance shops
14. Data processing	2. Food service	7. Air compressors
	a. Kitchen	8. Scrap collection area
	b. Dining	9. Vehicle storage
B. Production	c. Vending machines	10. Fire protection
1. Industrial engineering	3. Lavatory	a. Extinguishers
2. Production control	4. Smoking area	b. Hoses
3. Quality control	5. Lounge area	c. Equipment
a. Receiving inspection	6. Recreation area	d. Sprinkler valves
b. In-process (floor)	7. Parking	11. Stairways
c. Final inspection	8. Time clock	12. Elevators
4. Plant engineering	a. Bulletin boards	13. Plant protection

functions, etc. Associated with production activities are quality control, production control, plant engineering, storage and warehousing, shipping, activities, etc. All ancillary activities have to be identified in this step and their requirements as they relate to the design of the plant determined.

4.2.11 Determine Storage Requirements

A number of the requirements associated with raw material and product storage have already been considered to some extent in several previous steps; however, at this point in the planning exericse all the storage requirements need to be identified. For each of the proposed storage areas, volume and other requirements need to be computed.

4.2.12 Plan Service and Auxiliary Activities

In this step, the service-related activities identified in step 10 and the auxiliary functions are evaluated more carefully to determine the extent to which they will be needed and the requirements they will impose on the design of the proposed facility. Health and medical facilities, food service, locker room and lavatory requirements, etc., are evaluated in this step. Further, the office area and auxiliary office services needed to support the administrative activities are determined. The type of service activities required can be determined from preliminary estimates of the number of personnel required to operate the plant and the expected size of the facility. Information about the anticipated number of employees and job functions can be used with available tables to determine office space requirements, locker room and lavatory needs, health and medical facilities, as well as food services. Parking lot requirements, landscaping needs and environmental controls also can be evaluated by this procedure.

4.2.13 Determine Space Requirement

Steps 1 through 12 provide for the development of a flow plan, the identification and description of the required equipment and the identification of work areas, required services and auxiliary activities. The data generated in these steps provide the basis for establishing space requirements for the plant. The space needed for each item of equipment and for the operations to be performed can be determined from equipment specifications and manufacturers' literature. Space requirements for the ancillary equipment and operations also can be determined from equipment manufacturers' information. These space needs, combined with storage requirements developed in step 11 and the space anticipated for service and auxiliary activities, provide the first-cut estimate of the total space needed for the proposed plant. However, it should be noted that this rough estimate will be modified as more data are acquired from the successive steps in the design project.

4.2.14 Allocate Activity Areas to Total Space

This step is a continuation of the procedure used for determining space requirements. Area templates cut to scale are prepared for each of the identified activities and then arranged to facilitate the most effective relationship between activities. By this process, waste space is minimized, adequate space is ensured for all plant activities and none of the activities required are overlooked in the space allocation process.

4.2.15. Consider Building Types

In this step, building and construction requirements to meet the needs of the plant layout are considered. This phase of the design process may be best accomplished with the assistance of an architect. A discussion of the selection of building configuration, construction materials, location and building services is beyond the scope of this text.

4.2.16 Construct the Master Layout

Construction of the master layout is the integration of all previous steps and can be considered to be the culmination of the design effort resulting in the generation of the final product. The final layout is constructed with the aid of templates, usually to a scale of ¼ inch equals 1 foot. The first task in this step is to revise and adjust the original flow plan and diagrams to better accommodate the integration of the individual work areas, auxiliary activities and services. The integration of the individual work areas should facilitate the development of a comprehensive material flow plan (Figure 4-3).

To facilitate effective processing operations, a variety of different flow paths can be employed in the individual work areas. Straight line, zigzag, U shape, circular as well as odd angle flow patterns could be used for an individual work area. In this step, different flow paths are reviewed and modified to provide an integrated flow pattern to maximize the efficiency at the individual work areas, as well as for movement of materials. The conveyors, lift trucks, pneumatic tubes, etc., identified for use also would be reevaluated at this time to determine whether their operating schedules and requirements are compatible with each other when integrated into the plant layout. The operation of the major items of production equipment also are reevaluated at this point, and the final selection of material handling equipment to convey material between processing equipment is made.

The elevation at which the different operations and activities are to take place also must be considered at this time. As shown in Figure 4-4, there are six levels of plant production activity that should be considered in determining elevation requirements, process efficiency, flexibility and economy of space. Equipment operating requirements will frequently dictate the location of the material handling equipment and the location for supporting or auxiliary equipment. HVAC equipment, electric power lines, water, air and sewer lines are usually located either below or above the working floor to prevent their interference with the production processes.

With the development of specifications for the location of the material handling facilities, careful consideration also must be given to the location of aisles and passageways in the development of the master layout (floor plan). Determination of aisle width and length and the interrelationship between aisles are all extremely important for obtaining a smooth-running plant operation. The aisle serves as an orderly system for allowing personnel movement, conveyance of material, access routes for equipment repair and effective movement of safety equipment (firefighting equipment, cameras, etc.). The interrelationship between aisles and both the anticipated traffic plan for the facility and the anticipated location of building columns need to be considered in developing an effective floor plan.

Once the major processing equipment, material handling facilities, aisles and columns have been identified and located in the floor plan, the required storage areas can be incorporated. The service and auxiliary activities identified in steps 10 and 12 for administration, production, personnel and physical plant also can be incorporated into the master layout. Requirements for heating, ventilating, air conditioning equipment power lines and controls, maintenance and quality control operations, food service, lavatories, administrative offices, etc. are again reevaluated and the final locations for these activities established in the master layout.

The incorporation of all these items further identifies the design requirements for the building needed to house the processing plant. In the building design phase, costs, structural needs, selection of building materials, location of stairways, elevators, selection of environmental control systems, landscaping and building orientation (to take maximum advantage of transportation, sunlight and other climatic features) are all evaluated and resolved. The final design of the parking facilities also will be established in this phase of the master layout step. A detailed discussion of the technology associated with this task is beyond the scope of the book and more appropriately is left for architectural and design engineering texts.

Concurrent with the establishment of the master layout is the development of charts itemizing all equipment and material requirements for construction of the plant and the preparation of an operations table to define employment needs and operating schedules for the proposed facility.

4.2.17 Evaluate, Adjust and Check Layout with Appropriate Personnel/Obtain Approvals/Install Layout/Follow Up on Implementation of Layout

These final four steps have been combined because they are the final procedures necessary to take the master layout (floor plan) and convert it to an effectively operating production facility. The specific tasks necessary for reevaluating, checking and acquiring final approval for the plant

92 DESIGN PRINCIPLES

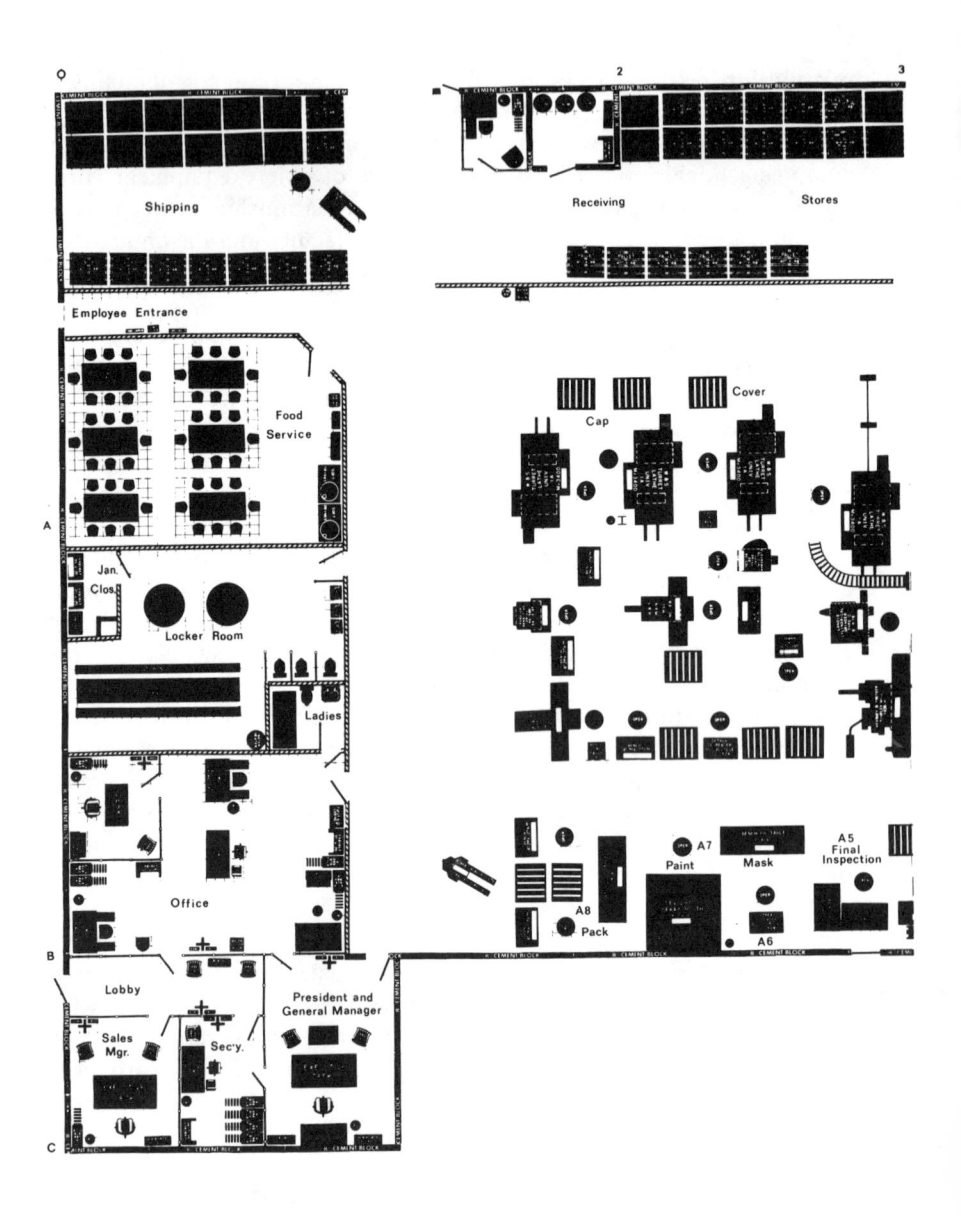

layout will depend on the mechanics of the individual firm engaged in the design project. However, to continually assure that development of the plant design takes advantage of the most recent information (technology, market conditions, regulatory requirements, etc.), continual review, reevaluation and administrative approval must be pursued. Once the

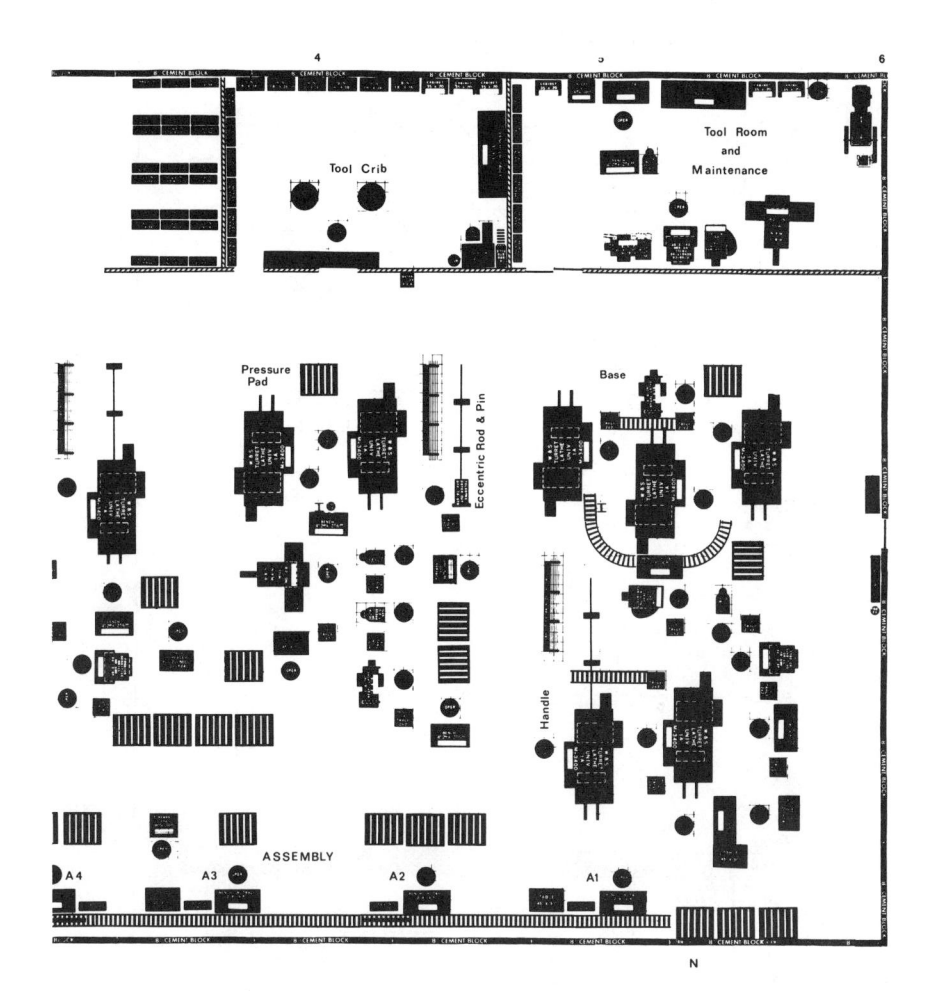

Figure 4-3. Finished layout [83].

plant is constructed, a shakedown process will be required for testing and possibly modifying the production operations.

It is important to recognize that the procedures outlined are meant to provide a general guideline for the design of a plant; however, not all will be needed, and a clear distinction between procedures will not always be

94 DESIGN PRINCIPLES

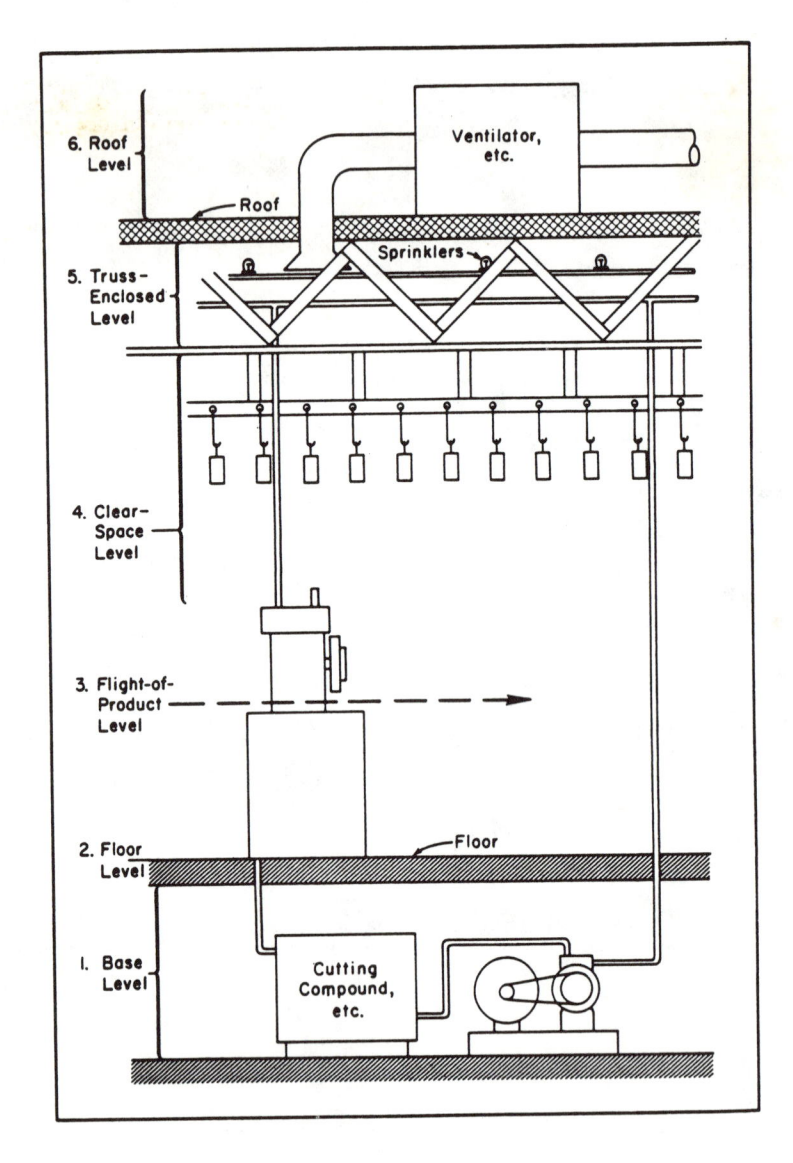

Figure 4-4. Levels of design in an industrial plant [83].

possible. Frequently, there will be a great deal of skipping back and forth between steps to develop an effective design. In addition, a variety of charts, diagrams and data sheets are available that have been developed to facilitate the design process. These data forms are discussed extensively by Apple [83] and other texts on the subject of plant design.

4.3 APPLICATION OF DESIGN PROCEDURES FOR RESOURCE RECOVERY FACILITIES

The sequence of procedure described and outlined in Section 4.2 can be effectively employed in the design of resource recovery plants. As would be expected, the application of these procedures to any specific industry will require the use of information and conditions specific to that industry. In addition, to better accommodate the unique requirements of each industry, a certain amount of modification would be needed to tailor the general procedures developed so they can be applied more effectively to the design process. The special information and procedure modifications appropriate for the design for resource recovery plants is discussed in this section.

4.3.1 Collect and Analyze Data

The data required for the design of a resource recovery plant is, to some extent, provided in Chapters 1–3 (an overview of the current state-of-the-art, general information about the nature and composition of municipal solid waste and unit operations).

In addition to the information provided, a considerable amount of site-specific data and equipment manufacturers' data also would be required for the development of an effective plant design. Site-specific information required includes maps describing the proposed site, a detailed description of the nature and composition of the waste, the market requirements, local codes, and EPA/OSHA regulations. The market data should include a comprehensive description of the products, the delivery cycle and the proposed default penalties. The American Society for Testing and Materials (ASTM), the National Center for Resource Recovery [84] and other organizations have prepared specifications for the various products available from municipal solid waste. Further, detailed descriptions and specifications of the equipment available for solid waste processing can be obtained from the various manufacturers.

4.3.2 Select Process

The analysis procedure serves as the initial step in identifying the production processes required. With the market requirements established and a comprehensive understanding of the waste stream, it is not too difficult to identify the processes needed to recover the required refuse-derived products. Unfortunately, in some of the resource recovery programs the design starts with a processing system already selected, rather than letting the market demands and waste characteristics determine the required processes.

Identification of the processes required to convert the raw waste materials to the final products is a critical phase. It requires that careful consideration be given to the processing technology (unit operations) available and the end product demands of the market. For example, the demand for energy in the form of steam can be effectively satisfied by mass burning of municipal refuse in a water wall incinerator. However, it could also be satisfied by a different sequence of unit operation processes, such as the production of fuel pellets (D-RDF) for combustion in an existing spreader stoker boiler or the use of a pyrolysis process to convert the municipal solid waste to a low-Btu gas, which could be used to fuel a gas-fired package boiler.

4.3.3 The Flow Plan

The sequence of unit operations selected for any resource recovery system should be based on an analysis of all the factors discussed above to determine the processing operations best suited to the specific resource recovery project. Once the required processes have been identified, a flow plan can be established. This flow plan should describe the logical sequencing of the required processing operations (Figure 4-1).

In most cases the first processing operation for consideration in a resource recovery plant is the receiving station for the solid waste. As the waste is received, it can go directly to storage, be subjected to some preliminary sorting, for example, to eliminate white goods (washing machines, stoves, etc.) oversized bulky wastes (sofas, frames, etc.), hazardous wastes (explosives, toxic materials), etc., or be conveyed to the next processing operation. In the mass burning operation, the raw refuse would be transported directly into an incinerator unit. A surge hopper most likely would be used so that the waste can be uniformly introduced into the unit. In a RDF recovery process, the refuse might be transported to a hammermill for shredding or to a trommel for size separation. If it is sent to a trommel first, then only the oversized material might be transported to a hammermill for size reduction. Once the material is the proper size, it might be transported to ferrous metal separation or to air classification to isolate the fuel fraction for recovery. Here again, the nature of the material and the market requirements will dictate the processing procedures followed for obtaining the required refuse-derived products. A compilation of typical flow diagrams for the major resource recovery processes have been described in Chapter 1.

4.3.4 Identify Material Handling Process

In working through the development of the flow plan, many variations must be tried before the final plan is established. Once a plan is finalized, the material handling processes to be used to transport the material between operations can be identified. The transport equipment selected must be compatible with the processing requirements identified. A detailed description of material handling equipment for resource recovery is presented in Chapter 3. A companion step with the identification of the general material handling requirements is the identification of the equipment to be used for each processing operation.

4.3.5 Select Equipment

Identification of the specific requirements for the equipment and the material handling facilities will require the establishment of an effective material balance. From the data bank assembled, the composition and characteristics of the solid waste should be fairly well established. Market requirements also have been determined so that the quantity of waste to be processed is known. For example, if the market requirement is for 50,000 lb/hr, of steam, 24 hr/day, and a water wall incinerator unit is to be selected, then the daily tonnage needed for processing can be calculated using the rule of thumb that 3 lb of steam are produced from 1 lb of refuse. The calculation for daily input tonnage is

$$50{,}000 \text{ lb steam/hr} \div 3 \text{ lb steam/lb refuse} \div 2000 \text{ lb/ton} \times 24 \text{ hr/day}$$
$$= 200 \text{ ton/day refuse}$$

If, instead of using a water wall incinerator, a modular incinerator were selected, then 2.2 lb steam/lb of refuse would be used in the calculation to determine daily tonnage requirements. If the market is for 500 ton/day of D-RDF (pellets), then a fuel recovery efficiency of 50% would be assumed and the following calculation would be used for determining daily tonnage:

$$500 \text{ ton/day} \div 0.5 = 1000 \text{ tons municipal refuse}$$

Once the required daily tonnage is established, the average quantities of different waste components can be assessed using Table 2-4. A facility processing 1000 ton/day of municipal solid waste would have to handle on a nominal basis 380 tons of paper, 35 tons of wood, 15 tons of textiles, 140 tons of yard waste, 160 tons of food waste, 25 tons of rubber and leather, 30 tons of plastic, 100 tons of metal, 100 tons of glass and 15 tons of other miscellaneous waste. However, it is important to remember that this is only an average; on a daily basis the quantity of each component is likely to vary considerably. In addition, there is also likely to be considerable variation due to seasonal effects. However, for initial planning purposes this composition will serve as an acceptable base.

The average bulk density of the raw refuse is about 220 lb/yd^3 and the moisture content will be about 25%. Using this information in combination with equipment performance data, it is possible to estimate how the refuse will be altered as it is carried through the various processing operations proposed in the flow plan. It should be noted that the composition described does not include the oversized, bulky waste or hazardous waste found in incoming municipal refuse because it is likely to be separated prior to processing. From the information obtained in the data collection phase, it should be possible to calculate the distribution of components resulting from the different processing operations. For example, if the first processing step is a trommel with 5-in. openings, about 546 tons would be recovered through the openings and 545 at the exit end of the trommel. From the data available in Figure 2-2, the distribution of waste components plus and minus 5-inches (assuming 90% screening efficiency) is tabulated in Table 4-5. Similar procedures for determining the distribution of the waste components can be employed for the different processing operations described in the flow plan. The end result of this procedure is the material balance (Table 4-3), which serves as the initial basis for calculating the equipment requirements in the flow plan.

Table 4-5. Component Distribution Recovered with a Trommel

Component	Percentage Dropping 5-in. Opening
Cardboard	10.8
Paper	45
Plastic	55.8
Textile	72
Ferrous	76.5
Nonferrous	86.4
Wood	76.5
Food	76.5
Garden	79.2
Sand and Rock	87.3
Glass	90
Composite	55.8

In addition to the material balance, it is also necessary to determine the number of processing lines and the hours of operation. For a mass-burning incinerator, it would be 24 hr/day, 7 day/wk. However, for a RDF system, only 8 or 16 hr/day may be required for processing and 4-8 hr/day for plant maintenance and cleanup. The number of processing lines to be employed is usually determined by balancing capital investment requirements and plant flexibility with equipment capabilities. If equipment capacities are available, a single processing line is probably the least costly and the simplest. However, a single line drastically reduces operating flexibility. Three processing lines tend to maximize operating flexibility and provide considerable redundancy and backup; however, three processing lines probably require extensive capital investment and complicated design operations. Therefore, two processing lines might prove to be an effective compromise.

In the operation of a resource recovery facility, the hours of operation and the number of processing lines are also affected by the solid waste collection and delivery schedule, the anticipated storage capacities and the market requirements. Resource recovery equipment is available in exess of 100-ton/hr capacities. A facility processing 1000 ton/day, 16 hr/day, with two processing lines would require equipment with processing capacities of 35 ton/hr. If the initial process is used to sort out part of the waste, then lower-capacity equipment would be required. Once tentative equipment sizing is accomplished, the process needs to be reevaluated to determine whether the equipment required is readily available and has a satisfactory record of performance at the needed capacity. This equipment selection process is one of continual trial and error until all the elements in the processing line are effectively determined.

4.3.6 Develop Work Station and Select Material Handling Equipment

To facilitate the detailed planning needed for the master layout plan, it is desirable to divide the plant activities into major work areas or stations. For most resource recovery design projects, this approach also will prove quite effective. The resource recovery process can, for the most part, be readily divided into distinct unit operations or work stations. These categories would include size reduction, size separation, density separation, ferrous recovery, RDF preparation, combustion, steam production, residue disposal and special treatments. Within each of these categories are distinct processing stations, which would be considered as individual work areas. For example, in a D-RDF production facility, using the flow

100 DESIGN PRINCIPLES

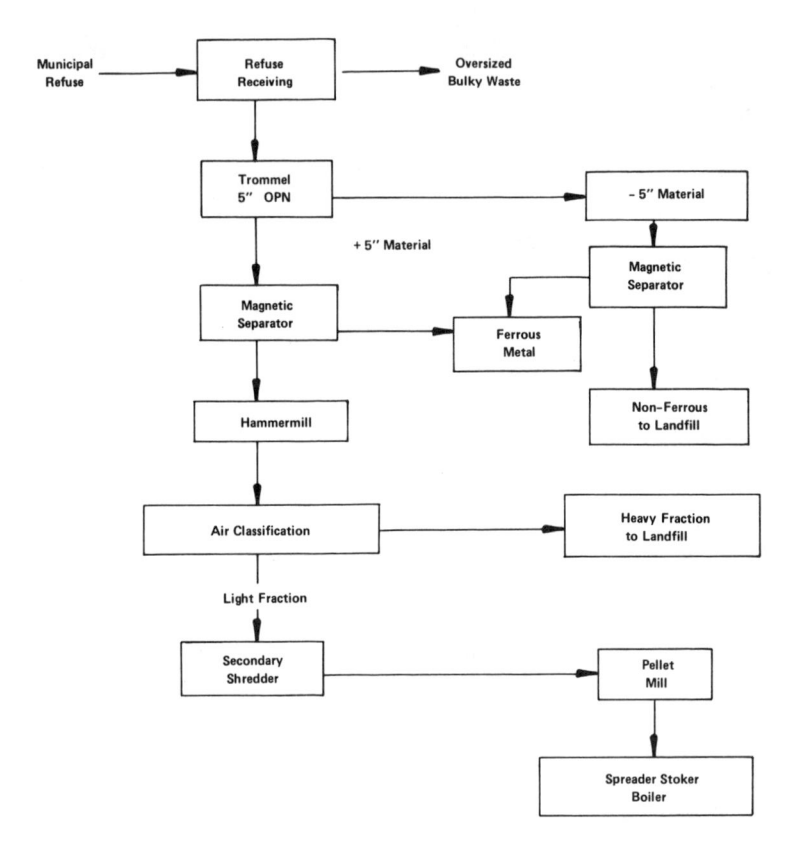

Figure 4-5. Flow plan for D-RDF plant.

plan shown in Figure 4-5, the following work areas can be identified: receiving, trommeling (primarily size separation), magnetic separation (ferrous recovery), shredding (size reduction), air classification (density separation), secondary shredding (size reduction), pellet mill (RDF preparation) and a spreader stoker boiler (combustion).

If the proposed system is to have more than one processing line, each operation in the processing line would be considered as an individual work station. Once each of the individual work station plans is completed, the specific material handling equipment necessary to transport the material between operations can be determined. Transport equipment selection and design are described in Chapter 4 in great detail and need not be discussed further in this section.

4.3.7 Coordinate Groups of Related Operations

To facilitate the material handling plan and the detailed layout, it is important to coordinate related operations—those requiring similar processing procedures, operations sharing auxiliary facilities, those using the same operators, those processes that have to be located adjacent to each other because of the sequence in the flow plan, and those operations that cannot be located close together for safety, environmental or other reasons.

An example of related operations in an RDF facility are the air classification process and the secondary shredding process for the production of RDF fluff. In this type of facility plan the air classifier and the secondary shredder are sequential processes, where the air-classified lights are transported into the secondary shredder. By reviewing the requirements of the two processes, one observes that it is possible to completely omit the need for a cyclone to deair the light fraction prior to shredding by exhausting the lights directly into a hammermill because they are sequential processes. This would reduce baghouse requirements for air pollution, eliminate the cost of a cyclone and reduce dust in the plant because the system would be closed. The hammermill ingests large quantities of air, and it appeared practical to use the air from the classifier for this purpose.

This example illustrates the opportunities available for more efficient operations by careful evaluation of coordinated groups in the design of resource recovery facilities. Other operations that have to be coordinated are those like the hammermills, which create serious vibrations in the facility and require excessive power for operation. With a better understanding of related operations, it may be possible to schedule the operation of energy-intensive units in a way that minimizes the electric load factor and results in lower utility costs.

4.3.8 Design Activity Interrelationships and Plan Service and Auxiliary Activities

The plant activities defined in Section 4.2.10 are, for the most part, similar to those required for most types of resource recovery systems. Administrative functions needed to manage the facility require accounting and data processing services, as well as purchasing and personnel management. In addition, most plants will require engineering activities for product control, quality control, receiving, storage, maintenance and process supervision. The personnel-related activities required will include health and medical facilities, food service, lavatories, parking, billboards,

102 DESIGN PRINCIPLES

fire escapes, drinking fountains and telephones. The physical plant-associated activities for a resource recovery facility would require HVAC units, power station, telephone/equipment room, air compressors, vehicle maintenance, fire protection, explosion protection, air, water and noise pollution control systems, stairways/elevators, and plant protection systems. Due to the nature of the work, it is important to have adequate lavatory and shower facilities where the workers can clean off dirt acquired during the work shift.

Of primary importance to the resource recovery facility are the fire and explosion control systems and the environmental control systems. Careful precautions must be taken to protect the plant against fires and explosions. Similarly, it is extremely important that the plant be a good neighbor and avoid the release of odors and other contaminants to the air and water surrounding the facility. In all likelihood, the facility should operate under negative pressure, and to facilitate these requirements the plant should conform to OSHA and EPA requirements. (These matters are discussed more fully in Chapter 5.)

4.3.9 Determine Storage and Space Requirements and Allocate Activities to Total Space

The general procedures described for determining space requirements also apply to the design of resource recovery facilities. Once the equipment requirements are established, the space needed for the processing equipment can be determined. Similarly, space requirements for auxiliary facilities and activities associated with the different processing steps also can be identified. Further, after most of the equipment and auxiliary facility space is established, support requirements can be determined and the needed space can be assessed. As with any other type of processing facility, adequate space must be allowed for personnel, as well as material movement. Adequate space allowances also must be made for maintenance operations and safety systems. Frequent material handling problems are likely to occur in the processing of MSW, and adequate provisions must be made to allow for equipment repair and replacement. Special allowances also must be made for noise and dust control systems. In addition to conventional concerns about health and safety requirements, it is particularly important to provide adequate space for fire and explosion control systems. In addition, adequate space must be provided for housecleaning operations. Any facility concerned with the processing of municipal waste must take special measures to maintain a high level of cleanliness.

Storage space requirements for resource recovery facilities also will require special considerations. Because of the nature of municipal solid waste, first-in, first-out practices must be followed. Because of putrefaction and corrosion problems, extensive storage capabilities are not advisable. For most resource recovery facilities three types of storage are necessary: (1) storage for municipal solid waste when it enters the plant, (2) storage for the product prior to leaving the facilities, and (3) storage for surges in various processing steps. Storage for receiving the municipal refuse when it enters the plant should not exceed one to two days. Landfill close to the facilities should be developed to serve as a backup in case the plant has to be closed for more than two days or there is an unusually heavy load of municipal refuse due to seasonal conditions. Space requirements for MSW can be determined by the density and required tonnage (ton/day, etc.) to be handled. Density values for MSW are presented in Chapter 2 (Table 2-5). A discussion of the storage facilities available for MSW is described in Chapter 3 (Section 3.4.2).

4.3.10 Construct the Master Layout/Establish Building Requirements

The procedures outlined in Section 4.2.16 for preparing the master layout are also appropriate for the design of resource recovery facilities. The layout should facilitate the effective recovery of materials and energy with maximum economy of space. Design of the buildings to house the facility and selection of the materials of construction usually are left to the discretion of the architectural and engineering firm selected for the project. However, for resource recovery systems, it is extremely important that the appearance be attractive to the public and that all odors and litter be contained within the facility and not be detectable outside the grounds. As mentioned, the plant probably should operate under negative pressure to prevent the escape of undesirable odors or dust. Although the selection of the building materials will be dictated by the requirements established by the architect, it is important that they be highly resistant to corrosive environments because acid vapors tend to be associated with many resource recovery operations.

4.3.11 Implement Tasks—Reevaluate and Approve Layout/Install Layout and Follow Up on Installation

Once the plant layout has been determined and the building requirements established, the next series of steps is directed toward implementing the plans established to bring the resource recovery facility to fruition. As most resource recovery plants usually are affiliated with local governments to one degree or another, the approval requirements are likely to be considerably different and more stringent than those found in the private sector. The project is most likely to be under great scrutiny and have more levels of review. The master layout and proposed construction plans that have been developed probably will have to be reviewed and approved by the engineering firm associated with the program, as well as by the engineering department of the local government. Further, the elected officials and the public sector will have to be briefed about the project plans and their approval acquired. It is especially important to have enthusiastic support for the program within the public sector, which can best be achieved by frequent public briefings throughout every stage of the resource recovery program.

4.4 SUMMARY

In this chapter some general rules for plant design were reviewed and their application to resource recovery systems described. The twenty basic steps for designing a plant were presented and discussed and the eight major objectives for effective plant design presented. Although resource recovery systems may impose some special problems, the basic procedures described for plant design can be effectively employed. However, it is important to recognize that the specific sequence of procedures followed for any particular plant design project will be dictated by specific circumstances of the project. Frequently the order of steps followed will be changed and several steps combined.

All basic plant design procedures have been described, and it is up to the resource recovery plant designer to employ them in the most effective way possible.

CHAPTER 5

PLANT DESIGN EXERCISE

5.1 INTRODUCTION

In 1974 AMAX Incorporated provided a grant to the University of Dayton to develop a detailed design and economic analysis of a resource recovery system to serve the 600,000 residents of an urban county in Ohio [85]. Based on an assessment of the solid waste management needs for the community, it was determined that a dry separation system would be most compatible for the community needs and the available long-term energy market. The proposed system would recover the organic fraction for use as a fuel in a utility boiler, the ferrous metal for sale to the steel industry and the remaining inorganic mineral material (glass, crockery, stone, etc.) for use as fill. The results of this program have been selected to serve as an example to demonstrate the procedures required for the design of a resource recovery system.

In this study, the first step was the development of a flow plan based on the best available commercial technology. From this flow plan, a detailed material balance for each stage of the process was developed. The information from the material balance served as the basis for equipment selection and the development of the architectural plan for the proposed facility. The design developed can serve as a model for evaluating the design parameters for a resource recovery facility. However, it should be pointed out that several resource recovery facilities have been built and put into operation since this study was conducted. The operating experience from these facilities provides a number of new insights into the design requirements for resource recovery systems. In addition, a number of laboratory and pilot studies to advance the state-of-the-art also have been initiated.

In this plant design example, the sequence of steps followed was altered to meet the circumstances of the design project. The following sequence was followed:

1. Flow plan development
2. Material balance development
3. Equipment sizing and selection
4. Plant layout
5. Development of building and facility requirements

However, it should be observed that step 1 is a combination of Apple's steps 1 through 4; steps 2 and 3 have replaced Apple's steps 5 and 6; step 4 is a combination of Apple's steps 7 through 11, 13, 14 and 16 through 20; and step 5 is a combination of Apple's steps 12 and 15. This example is included to illustrate the flexibility of the plant design procedures presented.

5.2. COMMUNITY DESCRIPTION AND REQUIREMENTS

The model county is an urbanized community of 600,000 residents located in Southwest Ohio. It was estimated that the county generated between 300,000 and 350,000 tons of municipal solid waste annually, while the county's two incinerators processed 250,000 tons of refuse. It was assumed that the resource recovery facility would be designed to process all the waste generated within the County (350,000 ton/yr). The proposed facility would receive refuse 24 hr/day, 7 day/wk and would process refuse 16 hr/day, 5 day/wk. Equipment maintenance and plant cleanup would be carried out during the other 8 hr/day, 5 day/wk. During the three weekend shifts, only maintenance, housekeeping and refuse-receiving functions would be performed. As the refuse generation rate is seasonal, an average monthly rate was not established; however, based on county data, a schedule for projected monthly distribution was developed. This schedule is presented in Table 5-1. The county's refuse was analyzed to determine the composition of the waste. The results of this analysis are presented in Table 5-2.

5.3 FLOW PLAN

The development of a flow plan to provide an effective resource recovery system consistent with the most advanced proven technology available was the first step in this plant design project. During this part of

Table 5-1. Projected Refuse Delivery Schedule

Month	Percentage	Average Monthly Tonnage	Work Days per Month	Projected Daily Tonnage
January	6.2	21,700	22	986
February	5.8	20,300	20	1,015
March	6.9	24,150	21	1,150
April	8.9	31,150	21	1,483
May	10.2	35,700	22	1,623
June	10.0	35,000	20	1,750
July	8.8	30,800	22	1,400
August	8.8	30,800	22	1,400
September	9.2	32,200	20	1,610
October	8.7	30,450	23	1,324
November	8.9	31,150	19	1,639
December	7.6	26,600	20	1,330

Table 5-2. Model County Refuse Composition

Component	Range (wt %)	Average (wt %)
Food	5–30	20
Yard	5–30	
Paper	30–60	47
Glass	5–15	7
Metal	5–15	10
Plastics	0.5–4.0	
Textiles	0.5–5.0	11
Leather and Rubber	0.5–4.0	
Wood	0.5–7.5	2
Miscellaneous	0.5–5.0	3

the project, the end products desired were identified and the unit operations available to process the refuse to obtain these products were determined. Using trial and error procedures, the best plan for coupling the different unit operations required was determined. The final flow plan developed for the project is presented in Figure 5-1. Those processes shown by the dotted line are presented as alternatives or additional processes that could be added at a later date. The data used in compiling this flow plan were based on information from the National Center for Resource Recovery, the University of Dayton, study, the Bureau of Mines pilot plant studies, the St. Louis demonstration plant and manufacturers of resource recovery equipment. The major product of this process is a

108 DESIGN PRINCIPLES

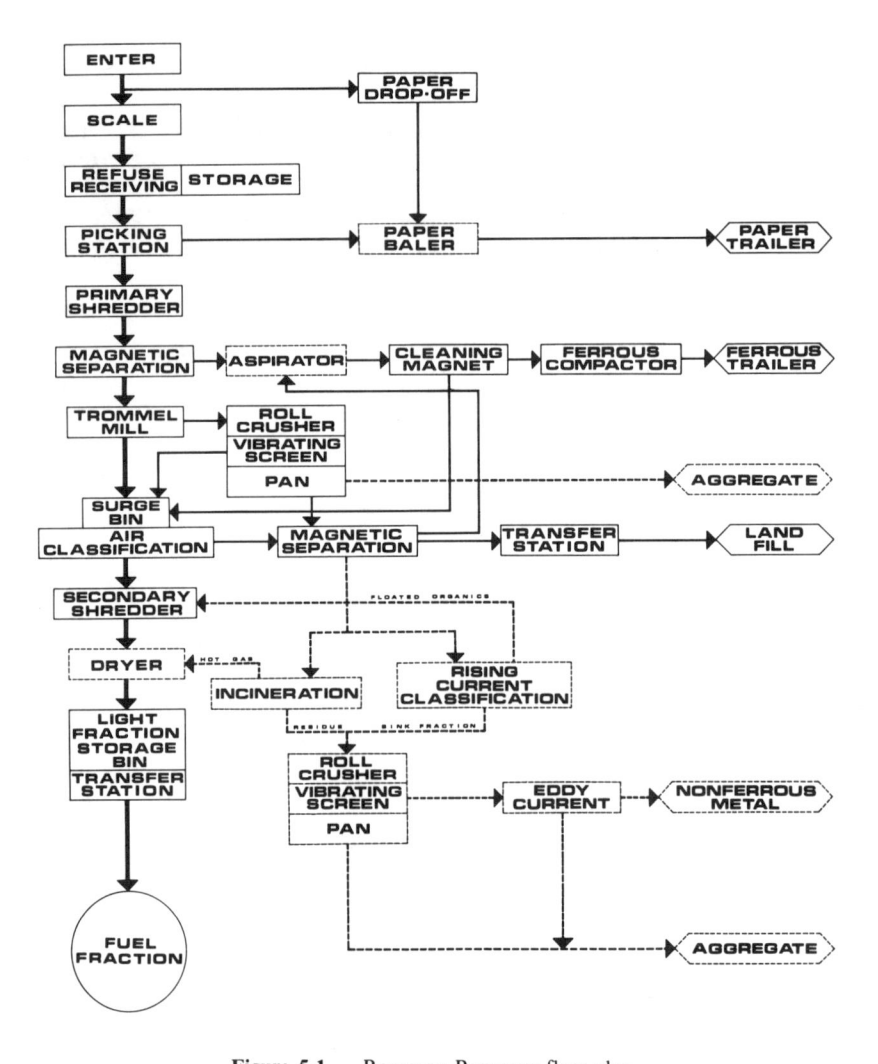

Figure 5-1. Resource Recovery flow plan.

refuse-derived fuel, obtained by shredding the refuse and removing the metal, glass and other undesirable inorganic materials. Two secondary products are also planned for recovery: ferrous metal and a sand aggregate.

As shown in the flow plan, the refuse trucks enter the facility, are weighed in and directed to the receiving area. The refuse is deposited on a tipping floor for processing into the system. Front-end loaders move the refuse to apron conveyors for transport past picking stations to a primary shredder. Handpicking stations have been located in front of the shredder unit to permit manual recovery of newsprint and possibly corrugated paper. At these stations the refuse going into the shredder also can be monitored to ensure that hardened steel objects and hazardous items (explosives, etc.) are not fed to the shredder. At a belt speed of 30–40 ft/min (6–12 in. bed depth), a picker should be able to sort from 800 to 1500 pounds of newsprint from the refuse stream. Two pickers per line (one on each side) could recover about 1 ton/hr of paper. This manual sorting operation would be maintained only during those periods in which the market price of paper justifies the process, and the picked newsprint and corrugated paper then would be conveyed to a trailer for shipment to a paper mill. An automatic baler for compacting the paper into a bale 5 ft × 3-½ ft × 3-½ ft (average density 30 lb/ft^3) can be added at a later time if necessary.

Horizontal hammermills have been selected for the proposed plant. These mills must be capable of processing both the normal municipal refuse (residential, commercial, institutional) and occasional bulky items, such as household appliances (washing machines, refrigerators, etc.), demolition waste (2 × 4's, concrete rubble, etc.), automobile tires, water heaters, tree limbs, furniture, etc. These primary shredders should have a nominal capacity of 45 ton/hr and reduce all refuse to less than 6 in. with 75% less than 4 in. The hammermills also should have the ability to reject (without being damaged) unshreddable items such as hardened steel shafts and gears. The bulk density of the shredded refuse produced will be between 5 and 20 lb/ft^3 and average about 10 lb/ft^3. The shredded refuse is discharged onto a vibrating pan conveyor, which uniformly feeds the shredded refuse to a belt conveyor going to the magnetic separator.

Two types of magnetic separators were considered: an overhead suspended belt magnet and a drum magnet. The magnet manufacturers recommended that either the overhead belt or drum magnet be suspended over the discharge of the conveyor within 8–12 in. of the flowing refuse. They further recommended that the refuse conveyor be traveling at speeds of 200–400 ft/min (average 300 ft/min). At a belt speed of 300 ft/min, the refuse would be 1-½ in. deep on the 6-foot-wide conveyor. For this proposed plant, a drum magnet was recommended for the initial separation phase. It is estimated that about 83% of the ferrous scrap in the refuse would be recovered at this primary separation stage and that about 12% organic contamination would be included with the recovered ferrous fraction. In addition, some 3% nonferrous metal is also likely to be in the ferrous fraction.

The ferrous fraction recovered is deposited onto a common conveyor to a secondary magnetic separation process (cleaning magnet) and its density is about 35 lb/ft³. The nonferrous material is conveyed from the magnetic separator into a trommel mill (rotary screen) to remove the fines from the nonferrous refuse.

The fine glass, ceramic and stone particles enmeshed in the shredded paper and plastic materials tend to get carried with the air-classified light fraction that is to be used for fuel. These particles, when in the fuel fraction, cause erosion problems in the pneumatic transport system and add to the ash and slagging problems in the boiler. It is desirable, therefore, to remove these fines from the shredded refuse. It should be noted that almost all the glass, ceramic and stone materials are less than ½ in. in size after primary shredding and more than 75% is less than ¼ in. in size. A rotary screen, 30–35 ft long with ½-in. wide openings, should facilitate the removal of these fines. A vibrating flat screen also could be used for this application; however, it is more susceptible to jamming. The particle size analysis data indicate that 15–25% (average 20%) of the shredded refuse can be less than ½ in. in size. Assuming a screening efficiency of 75%, it is likely that about 15% of the shredded refuse would be found in the fine fraction collected. Based on the average refuse distribution, it is estimated that about 45% of the fines would be organic, 36.7% would be glass, 15% dirt and stone and 3.3% metal. The density of the recovered fines is about 58 lb/ft³. The material passing through the trommel is conveyed to a surge bin for introduction into an air classification system.

The estimated distribution of components in the fines separated by the trommel would include a large quantity of organic material (45% of the fines). As a major objective of the proposed facility is the recovery of organic material for use as fuel, it would be desirable to recover these organic fines to add to the fuel fraction. Water classification and possibly air classification could be used to separate the organic fines. Another possible way of recovering the organic material would be to pass all the fines through a roll crusher and then over a 15 mesh vibrating screen. The roll crusher should pulverize all the glass, stone and rubble, shred and compress most of the organics and elongate the metal. The dust, dirt, glass and stone would go through the 16 mesh screen, but the majority of organics and metal would stay on the 16 mesh screen. The organic and metal fractions recovered from the 16 mesh screen would be conveyed back into the processing system at the surge hopper for air classification. However, initially the fines (minus 16 mesh) are to be combined with the heavies for disposal at a landfill. It should be noted that these fines could be washed and used as a substitute for fine aggregate sand. Although this fine powder processing procedure has not been used extensively for raw refuse, it has been used successfully by the Bureau of Mines in processing incinerator residue and in several compost processes. It is felt that the roll crusher and screen would be an effective low-cost means for recovering the organic and metal fines.

As the organic fraction is the major product for recovery, the classification system selected should maximize the separation between the inorganic (glass, metal, stone, etc.) and organic (paper, wood, textiles, plastics, etc.) materials in the refuse. Based on the information available, a vertical zig-zag air classification unit was recommended for the proposed plant. Preliminary estimates in the proposed flow plan project that 85.6% of the refuse coming into the classifier will report to light fraction and 14.4% will drop out of the classifier into the heavies fraction.

The recovered light fraction is pneumatically conveyed to a cyclone for partial deairing and then dropped by an interconnecting chute into the top of a secondary shredder. As this light fraction is to be used as a fuel supplement, it should all be less than 1½ in. in size with 75% less than 1 in. in size. The density of the light fraction entering the secondary shredder is 6–7 lb/ft^3. A horizontal hammermill under negative draft was recommended for this process. From the hammermill, the shredded lights would be pneumatically conveyed to a second cyclone for deairing and then dropped into a storage bin. The lights from the storage bin would be conveyed to a transfer station for compaction and subsequent transfer to a boiler facility for use as a supplement fuel.

The material dropping to the bottom of the air classifier, termed heavies (30 lb/ft^3), will contain the heavy organics (wood, rubber, molded plastics, leather, wet paper, food wastes, etc.), as well as the metal, glass and stone in the shredded refuse. From the classifier, the heavies will be conveyed to a heavies magnetic separator. In addition, the minum 16 mesh fines would be added to the heavies conveyor going to the magnetic separator. A drum magnet was proposed for this secondary ferrous recovery process. This recovered ferrous fraction with trace contaminants would be conveyed to a cleaning magnet. The nonferrous heavies passing under the magnet will be conveyed to a transfer station, where they will be compacted into a trailer (59 lb/ft^3) for transport to a sanitary landfill.

The ferrous fraction from the primary magnet and the ferrous recovered from the heavies magnet are combined onto a common conveyor to the cleaning magnet. This final magnetic separation stage is designed to provide a cleaner ferrous fraction for recovery. The material is conveyed to the drum and the nonferrous materials drop past the drum to an outfeed conveyor. An aspiration system can be combined with the drum magnet as an option to pull off the light nonferrous fraction. The ferrous fraction picked up by the magnet is conveyed to a ferrous compactor for densification (minimill). The ferrous metal having a density of 35 lb/ft^3 is compacted to a ferrous processing plant, and the nonferrous fraction is conveyed to the air classifiers for further processing.

Although the proposed plan specifies the disposal of the heavies and minus 16 mesh fines in a sanitary landfill, several alternate means of handling these materials can be considered: wet classification (rising current classifier), of the heavies for further refuse recovery, heavy media separation, eddy current processes, dry screening and incineration. With the rising current classifier, most of the remaining organics in the heavies fraction can be floated off, leaving the glass, stone and metal. Some of the plastics and rubber and small bits of paper and cloth will remain in the sinks fraction. The residue sinks could be further processed in a roll crusher and through a vibrating screen (16 mesh). Through this procedure, the nonferrous metals can be separated and recovered from the glass and plastics in the sink fraction. The nonferrous metal will be elongated in the roll crusher and not pass through the 16 mesh screen. The glass, stone and molded plastics will be pulverized in the roll crusher and pass through the screen. This fine fraction could be added to the washed minus 16 mesh fraction from the rotary screen and used as an inert fill (aggregate). The organics floated off in the rising current classifier can be recovered, dewatered by filter pressing and added to the fuel fraction at the secondary shredder. The major objection to using the rising current classifier is the additional problems resulting from the use of a wet process.

Another process that could be used for handling the nonferrous heavies is incineration. In this process the organics would be burned, leaving the ash and noncombustible glass, stone and nonferrous metal. This residue could be processed in the roll crusher and through the 16 mesh screen to separate the nonferrous metal from the inert fill fraction. The hot gasses from incineration could be used to dry the refuse after shredding to improve the quality of the fuel fraction and aid in subsequent separation phases. A number of benefits would be obtained by drying the refuse.

A consolidated analysis of the flow plan developed is presented in Table 5-3.

5.4. MATERIAL BALANCE

The detailed material balance for each stage of the developed flow plan was established. The projected distribution for refuse components at each stage of the flow was based on expected behavior for an average refuse sample. However, the fluctuations in refuse, due to seasonal changes and the usual daily variations, will result in moisture and compositional changes that should alter behavior of waste products at each stage of the resource recovery process. The material balance developed is presented in Table 5-4. The starting values used for the different refuse components were taken from the refuse analysis for the model county. It should be recognized that the information developed in the mass balance would serve as the basis for determining the equipment requirements for the proposed plan.

5.5. PLANT DESIGN

It was estimated that the model county would generate 350,000 ton/yr of municipal refuse. To satisfactorily process this quantity of refuse 260 day/yr and allow for the necessary redundancy, it was decided that the proposed plan would contain three processing lines, each having the capacity to handle 45 ton/hr. when all three lines are in operation 16 hr/day, the plant would have the capacity to process 2160 ton/day.

Based on the experience obtained from the operating resource recovery facilities, it is clear that the proposed plant was overdesigned. It would have been more effective to have designed the plant with only two processing lines, each with a capacity of 50 ton/hr, giving the facility a capacity of 1600 ton/day. Any excess refuse or refuse that cannot be processed due to equipment breakdown should be disposed of in a nearby landfill associated with the operations of the resource recovery facility.

5.5.1 Equipment Selection

Using the flow plan that has been developed and the material balance that has been established, the equipment needed to implement the proposed process plan for a facility with the maximum processing capacity of 2116 ton/day was determined. Working with equipment manufacturers it was possible to determine the equipment specifications required for an integrated flow plan, which would permit smooth interfacing of the different items of equipment. Established vendors were asked to submit costs and descriptions for each of the items identified for the flow plan (see Table 5-5). The data obtained served as the basis for developing the plant layout.

5.5.2 Plant Layout

Utilizing the assembled information, a floor plan arrangement for the processing equipment was developed. The final floor plan arrangement was selected through a trial and error process, with the final arrangement

114 DESIGN PRINCIPLES

Table 5-3. Resource Recovery Flow Plan Analysis

Process Steps	Process Requirements	Material Description & Comments
I. Refuse Receiving		
A. Truck weight-in	Plant Entrance—accommodates 360 trucks/day; drop area—for bundled newsprint and truck crew rest area	Municipal refuse[a]—2160 ton/day Received 24 hr/day; peak times: 9:30–11:30 and 2:30–4:30
	Scale house—monitor refuse in and products out; truck scales—two of 50-ton capacity	Collection trucks—municipal, commercial and private vehicles
B. Truck unloading	Unloading stalls—twelve of 15-foot width; dumping area—a lower level work floor (average truck unloading time of 8 min or 90 trucks/hr); truck exit—lane for departing vehicles	Raw refuse—~8 lb/ft^3
C. Refuse storage	Lower level work floor, in addition to facilitating introduction of refuse into the processing system, is available for refuse storage; work floor capacity—about 432,000 ft^3 (one day's storage)	Raw refuse—loose ~6 lb/ft^3; in storage, 10–15 lb/ft^3 2160 ton/day (received in surges)
II. Process Initiation	Front end loaders—three to move refuse to storage or onto entrance conveyors. For 2160 ton/day plant—three lines capable of handling 45 ton/hr (187 ft^3/min/lin); entrance conveyors—four apron type, 12 ft wide by 20 ft long by 3 ft deep; belt speed—5–15 ft/min; refuse conveyors—three apron type, 6 ft wide speed of 35 ft/min	Raw refuse—averages ~8 lb/ft^3 Plant will process refuse 16 hr/day (2 shifts); third shift—plant maintenance and housekeeping Conveys refuse (depth of 1 ft) to picking station (30° incline), then, past the pickers (belt parallel to floor) and up to primary shredder (30° incline)
III. Handpicking of News-print	Platforms for pickers[b]—six platforms 6 ft long	Permits standing space for one picker/side;

PLANT DESIGN EXERCISE 115

and Corrugated (optional)	by 3 ft wide with discharge chute to paper outfeed conveyor	raw refuse, 8 lb/ft^3, 45 ton/hr/line
	Paper outfeed conveyor—one belt, 2 ft wide [paper baler—6–8 ton/hr capacity (automatic operating unit)—optional]	Conveyor carries picked paper to a trailer—newsprint and corrugated paper—1000 lb/picker, or 1 ton/hr line; giving ~48 ton/day, 15 lb/ft^3 (Baled paper, 30 lb/ft^3)
IV. Primary Shredding	Hammermills—three horizontal units, 45 ton/hr capacity; shredder infeed conveyors, three vibrating pan types, 6 ft wide (optional)	Shredder must be capable of handling bulky wastes and reduce all material to a nominal 6 in. with 75% minus 4 in.—shredded refuse, ~10 lb/ft^3 (range of 5 to 20)
	Outfeed conveyors—three vibrating pans, 80 in. wide, 1 in. stroke	Shredded refuse, minus 6 in., ~10 lb/ft^3
	Shredded refuse conveyors—three belt units, 5 ft wide, ~300 ft/min	
	Control Booth	Operator oversees and controls infeed and shredder operation
V. Primary Magnetic Separation	Drum magnets—three, 48 in. diameter, 7 ft wide	Shredded refuse drops from refuse conveyor past drum and onto conveyor to the trommel mill; ferrous picked up by drum.
	Ferrous conveyor—one, belt-type unit, 2 ft wide, ~300 ft/min	Conveys recovered ferrous to cleaning magnet; of the 3.94 ton/hr/line recovered, 85.7% is ferrous, 11.4% organic, and 2.9% nonferrous. Ferrous is mostly plus 1 in. cans, caps, nails, wire,cookware, hardware and appliances—density ~35 lbs/ft^3
	Nonferrous outfeed conveyors—three belt-type units, 4 ft wide	In feed to trommel mill, nonferrous shredded refuse 40 ton/hr/line, density ~7.5 lb/ft^3
VI. Fines Separation and Processing	Trommel mills (rotary screens)—three 6 ft wide by 35 ft long, ½ in. openings, on 12° incline	Nonferrous shredded refuse, 40 ton/hr/line; expected screening efficiency, 75%

116 DESIGN PRINCIPLES

Table 5-3. Continued

Process Steps	Process Requirements	Material Description & Comments
	Oversized material drops into surge bin, which feeds the air classifier	Plus ½ in., ~7.5 lb/ft³, ~33.5 ton/hr
	Fines conveyor—belt, 2 ft wide, minus ½ in. material to roll crusher	Fines (~20% of stream) are minus ½ in., ~58 lb/ft³, 6.7 tno/hr/line Fines include 15% dirt and stone; 36.7% glass; 3.3% bits of metals; 5% bits of rubber and plastic; 3.3% bits of wood; 16.7% finely shredded paper and 20% grass, seeds, bits of bone, etc.
	Roll crusher, 30 ton/hr capacity, feeds onto 16 mesh vibrating screen and pan	Rolls set to pulverize glass and stone to minus 16 mesh
		The glass and stone, being friable, would be in pan fraction; metals, being malleable, would be plus 16 mesh
		Most organics also in plus 16 mesh, fraction
VII. Air Classification	Surge bins (for air classifier)—three 300-450 ft³ capacity	Maintains uniform flow of refuse, ~37 ton/hr/line or ~190 ft³/min to air classifier Refuse includes plus ½ in. from trommel mill, plus 16 mesh from screening and nonferrous from cleaning magnet
	Air classifiers—three zig zag units, 45 ton/hr capacity	
	Lights conveyors—three pneumatic units	Convey lights to cyclones feeding secondary shredders; material, ~6.5 lb/ft³ and ~31.6 ton/hr/line is predominantly organic and includes paper, plastic, grass, leaves, wood, etc.
	Heavies conveyor—belt, 2 ft wide, 300 ft/min	Conveys air-classified heavies (30 lb/ft³, 15.7

PLANT DESIGN EXERCISE 117

VIII. Magnetic Separation of Heavies and Fines	Drum magnet, 36 in. diameter, 4 ft wide	ton/hr) to magnet With the minus 16 fines from the roll crusher added, the mix is ~59 lb/ft³, 27.3 ton/hr
	Ferrous outfeed conveyor, belt unit, 2 ft wide	Conveys ferrous to cleaning magnet
		Ferrous fraction and entrained contaminants ~35 lb/ft³, ~1.7 ton/hr
	Nonferrous outfeed conveyor, belt unit 2 ft wide	Nonferrous fraction ~25 lb/ft³, ~25.6 ton/hr Carries material to transfer station for transport to landfill
IX. Final Ferrous Processing	Drum magnet, 36 in. diameter, 4 ft wide	Feed is combination of ferrous fraction from primary magnet and ferrous fraction from magnet separation of heavies and fines Feed is ~ 35 lb/ft³ ~13.5 ton/hr
	Nonferrous conveyor—one belt, 2 ft wide	Conveys nonferrous (~20 lb/ft³ ~1.7 ton/hr), which has fallen past magnet, to air classifier surge bin
	Ferrous conveyor—one belt, 2 ft wide (aspirator and pneumatic conveyor optional)	Conveys ferrous (~35 lb/ft³ ~11.8 ton/hr), which has been pulled off by drum magnet, to minimill (aspirates light organic contaminants and conveys them to secondary shredder)
	Minimill (vertical nuggetizer)—one with capacity of 15 ton/hr	Densifies ferrous fraction to 70-75 lb/ft³; 11.8 ton/hr
	Minimill outfeed conveyor—one belt, 2 ft wide	Carries ferrous fraction to trailer for shipping
X. Secondary Shredding	Cyclones—three above secondary shredder (does not require airlock)	Lights ~6.5 lb/ft³, 31.6 ton/hr/line pneumatically conveyed from air classifier to cyclones Partially deaired lights from cyclones fall down a chute into the shredder

118 DESIGN PRINCIPLES

Table 5-3. Continued

Process Steps	Process Requirements	Material Description & Comments
	Hammermills—three horizontal, 35 ton/hr capacity	Air is recycled back to bottom of air classifier Reduce all material to minus 1½ in. and 75% to minus 1 in. Hammermill should be under negative draft and equippped with air pan to carry finely shredded refuse to cyclone on top of lights bin
	Pneumatic conveyors—three	Transports shredded refuse to lights storage bin
	Cyclones—three on storage bin	Finely shredded lights ~6.5 lb/ft^3
XI. Light Fraction Storage	Storage bins—three, each having 500 tons capacity	Each bin will have a transfer station built under the outfeed system Here the refuse will be compacted before shipping to the fuel consumer
	Air cleaning equipment—three, equipped with afterburner combustor for odor removal (baghouse dust collector)	Dust collected (0.3 ton/hr) is returned to heavies conveyor for disposal in landfill

[a] Municipal refuse—residential, commercial and institutional solid wastes: paper, 47%; food and yard waste, 20%; wood, 2%; plastic, leather, rubber textiles, 11%; glass, 7%; metal, 10% (ferrous, 9%); dirt, stone and rubble, 3%. (Data from Montgomery County Study.)
[b] Pickers also could sort other valuable materials (aluminum, etc.) pick out unshreddable or hazardous materials or cut open bagged refuse for inspection.

Table 5-4. Material Balance

Refuse Components (% as received)

Process	Paper	Yard & Food Wastes	Wood	Plastics, Rubber, Leather, Textiles	Total Organics	Ferrous Metal	Non-ferrous Metal	Glass	Dirt, Stone, Rubble, etc.	Total	Material Destination
1. Refuse Receiving	47	20	2	11	80	9	1	7	3	100	Refuse moved to entrance conveyor by front-end loader
											Conveyor moves refuse to picking station
2. Paper Picking	2.23	—	—	—	2.23	—	—	—	—	2.23	To trailer for shipment
3. Primary Shredding	44.77	20	2	11	77.77	9	1	7	3	97.77	After the picking station, the refuse conveyed to shredder; all material reduced to less than 6 in. and taken to magnet separation
4. Magnetic Separation											
(a) Ferrous fraction	0.5	0.25	—	0.25	1.00	7.5	0.25	—	—	8.75	Goes to cleaning magnet
(b) Nonferrous fraction	44.27	19.75	2	10.75	76.77	1.5	0.75	7	3	89.02	Goes to trommel mill
5. Trommel Mill											
(a) Fines	2.5	3.0	0.5	0.75	6.75	0.25	0.25	5.50	2.25	1.500	Goes to roll crusher
(b) Oversized	41.77	16.75	1.5	10.00	70.02	1.25	0.50	1.50	0.75	74.02	Goes to air classifier
6. Roll Crusher	2.5	3.0	0.5	0.75	6.75	0.25	0.25	5.50	2.25	15.00	Goes to vibrating screen
7. Vibrating Screen											
(a) =15 mesh deck	2.25	3.0	0.5	0.50	6.25	0.25	0.25	—	—	6.75	Goes to air classifier
(b) Pan (−16 mesh)	0.25	—	—	0.25	0.50	—	—	5.50	2.25	8.25	Goes to A.C. heavies magnet
8. Air Classifier	44.52	20.00	2.0	10.75	77.27	1.50	1.00	1.50	0.75	82.02	
(a) Lights	43.40	16.00	1.00	8.35	68.75	0.25	0.25	0.50	0.50	70.25	Goes to secondary shredder
(b) Heavies	1.12	4.00	1.00	2.40	8.52	1.25	0.75	1.00	0.25	11.77	Goes to A.C. heavies magnet
9. A.C. Heavies Magnet	1.37	4.00	1.00	2.65	9.02	1.25	0.75	6.50	2.50	20.02	
(a) Ferrous fraction	—	—	—	—	—	1.25	—	—	—	1.25	Goes to cleaning magnet
(b) Nonferrous fraction	1.37	4.00	1.00	2.65	9.02	—	0.75	6.50	2.50	18.77	Goes to sanitary landfill
10. Cleaning Magnet	0.5	0.25	—	0.25	1.00	8.75	0.25	—	—	10.00	
(a) Ferrous fraction	—	—	—	—	—	8.75	—	—	—	8.75	Goes to compactor and then to trailer
(b) Nonferrous fraction	0.5	0.25	—	0.25	1.00	—	0.25	—	—	1.25	Goes to air classifier
11. Secondary Shredder	43.40	16.00	1.00	8.35	68.75	0.25	0.25	0.50	0.50	70.25	Goes to lights storage bin
12. Storage Bin	43.40	16.00	1.00	8.35	68.75	0.25	0.25	0.50	0.50	70.25	Goes to boiler

120 DESIGN PRINCIPLES

Table 5-5. Equipment and Facilities[a]

Plant Area	Equipment	Manufacturers	Model
1. Plant Entrance and Grounds	— —	— —	— —
2. Truck Crew Rest Area 　Men's and women's lavatories, security office and supply closet 　Lounge	Bathroom fixtures Intercom Cleaning equipment Furniture Vending Machines Furniture	— —	— —
Paper drop off	Trailer (supplied by paper company)		
3. Scale House	Gate control system, two 50-ton truck scales Scale instrument (99,980 lb capacity) with digital display Operator's console card Data recording, printout and storage system	Toledo Scales and Systems	810-830 MT-Scale
4. Refuse Receiving Area 　Truck dumping zone (12 stalls)	Traffic lights	— —	— —
Work floor and storage area	Three front-end loaders Four 12 × 20 apron conveyors Two 6 × 48 apron conveyors Two 6 × 55 apron conveyors One 6 × 90 apron conveyors	Caterpillar Equipment Cindaco Cindaco Cindaco Cindaco	#950 with 7-½ yd bucket C1, C2, C2′, C3 C4, C6 C7, C8 C5
5. Refuse Processing Area 　Control and observation tower	Intercom system Electrical control system (for entire		

(15 ft × 32 ft)	processing line) furniture, traffic system)	—	—
Electrical power station	Substation, starter systems, etc. (motor control center)	—	—
Paper picking station	Three picking stations (metal frame)	—	—
	One 2 × 72 belt conveyor	Cindaco	C24
	One 2 × 30 belt conveyor	Cindaco	C36
Primary shredding	Three 45 ton/hr hammermills	Grundler (Williams)	60 × 84
	Three 6-2/3 × 20 vibrating pan conveyors	Cindaco	C9, C10, C11
	Three 6 × 15 belt conveyors	Cindaco	C12, C13, C14
Primary magnetic separations	Three 4D by 7W drum magnet	Eriez Magnetics (Dings Co.)	Type SR
	One 2 × 90 belt conveyor	Cindaco	C25
	Three 4 × 42 belt conveyors	Cindaco	C15, C16, C17
Fines separation	Three trommel mills (rotary screens)	Triple/S Dynamics	6 × 30
	One 2 × 73 belt conveyor	Cindaco	C26
	One 3-1/3 × 26 belt conveyor	Cindaco	C28
Fines processing	One 30 ton/hr roll crusher (40 in. by 36 in. double smooth rolls)	Guendler	40 × 36
	One 16 mesh vibrating screen (enclosed with pan)	Triple/S Dynamics	Series 1000H Overstrom
	One 2 × 33 belt conveyor	Cindaco	C29
	One 2 × 88 belt conveyor	Cindaco	C33
	One 2 × 6 chute conveyor (metal frame)	—	—
Air classification	Three air classification systems (surge bin, conveyor, leveller, fan, zig-zag column, cyclone and air recycle pipe)	Combustion Power Co. (Rader Pneumatics)	45 ton/hr
Heavies processing	One 2 × 64 and one 2 × 30 belt conveyor	Cindaco	C27, C37

Table 5-5. Continued

Plant Area	Equipment	Manufacturers	Model
Final ferrous processing	One 3D × 4W drum magnet	Eriez Magnetics (Dings Co.)	Type SR
	One 2 × 18 belt conveyor	Cindaco	C35
	One transfer station	The Heil Co.	HTP-1000
	One 2 × 110 belt conveyor	Cindaco	C30
	One 2 × 71 belt conveyor	Cindaco	C31
	One 3D by 4W drum magnet	Eriez Magnetics (Dings Co.)	Type SR
	One 2 × 24 and 2 × 12 belt conveyors	Cindaco	C32, C34
Secondary shredding	One 15 ton/hr minimill (grinder)	Carborundum Co. (Eidal)	Model 4000
	Three transport chutes (cyclone to shredder)	—	—
	Three 35 ton/hr hammermills (with air sweep exit pan, transport tube and fan)	Williams (Grundler)	#475
	Three cyclones (on storage bin)	Williams (Grundler)	—
	Three air cleaners (bag-type dust collectors)	Williams (Grundler)	—
Lights storage	Three 500-ton storage bins	Atlas Co.	17-5T
	Three transfer trailer systems (located under bins)	The Heil Co.	HTP-1000
6. Office			
Manager's office			
Secretary's office	Furniture	—	—
Conference room			
Clerical office (mil, billing, payroll and records)	Furniture, vault		
Supply room	Supplies and equipment		

Lunch room	Furniture
Quality control lab	Lab equipment
Quality control office	Furniture
Men's & women's lavatories	Fixtures
7. Maintenance Building	
Foremen's offices (3)	Furniture
Security offices	Furniture
Lunch Room	Furniture
Maintenance shop	Tools and machine equipment
Tool crib and storage	Tools and supplies
Locker room and showers	Fixtures
Men's and women's lavatories	Fixtures
Supply closet	Supplies

[a] The equipment and manufacturers identified in this table were taken from the University of Dayton AMAX study [58] and are presented for purposes of example only.

124 DESIGN PRINCIPLES

being the one most compatible to the economy of space and interfacing requirement of the different items of equipment. A schematic of the final processing plant layout is presented in Figure 5-2. An elevation plan for the main processing line developed from the layout shown in Figure 5-2 is presented in Figure 5-3. Using the layout developed, it was possible to determine the space requirements for the processing system and to identify the major processing areas: (1) plant entrance and grounds; (2) truck crew rest area and bundle paper dropoff; (3) weigh station; (4) refuse receiving area; (5) refuse processing area; (6) maintenance building; and (7) administrative office area.

As stated, the proposed plan was designed for three processing lines having the capacity of handling 45 ton/hr (720 ton/day/line); thus, only two of the three available lines would be needed to operate for most months. The down line could be rotated for each shift. If trouble developed in an operating line, the out-of-service line could be reactivated. If, for some reason, the entire plant were shut down, one day's storage would be provided on the work floor. Additional storage space would also be available on the upper level tipping floor. The work floor would be used to accommodate surges and refuse delivery, should more than one line be out of service. During peak seasonal conditions, all three lines would be operated to prevent refuse pileup.

5.5.3 Refuse Receiving Plan

Refuse trucks entering the plant would stop first at the bundled paper dropoff and crew rest area. In this area there is a trailer for the bundled newsprint, operated by the paper collector, and a crew rest building. The crew rest building has a lavatory and lounge facility for the truck crew. After droppingoff the bundled paper and crew, the truck driver would continue on to the weigh station and then to the refuse dumping area. After dumping the refuse, the truck driver would circle back to the crew rest area, pick up his crew and then exit the plant.

The weigh station is used to control traffic into the plant. It weighs the incoming load, providing a printed record for both the driver and the data processing system. An automatic gate control system would be used to regulate the truck flow and prevent vehicles from entering the plant without weighing-in. An automatic manual recording system would be used to record the time, date and gross weight of the truck on the scale, along with the truck tareweight and identification from the plastic data card provided by the driver. Additional data such as billing code, truck route number, etc., also could be recorded. The information collected would be retained for billing and a receipt given to the driver. The driver would proceed to the designated tipping stall (controlled from an operations tower) and a new truck would be weighed-in. For trucks or private vehicles without plastic data processing cards, a manual recording system would be used. Trucks carrying the products recovered at the facility would exit through the weigh station so that they could also be weighed-out.

Refuse delivery at most plants occurs in surges. For the model county, peak refuse delivery to the disposal facility is from 9:30 to 11:30 A.M. and from 2:30 to 4:30 P.M. With this type of delivery pattern, the as-received refuse must be stored and then fed into the resource recovery system at a uniform rate.

Flow through or dumping turn-around time for a refuse truck usually varies from 5 to 10 minutes and will average about 8 minutes. This turn-around time extends from the time the truck enters the weigh station to the time it leaves the refuse dumping area. It was determined that 12 unloading stalls would be required for this proposed plan. This would allow the receiving facility to handle 90 trucks per hour at maximum capacity. With this arrangement, all 2160 tons could be handled during the four peak receiving hours if necessary. Refuse trucks entering the receiving facility would dump the refuse onto a lower level work floor (12 ft below the truck unloading floor). Front-end loaders then would be used to move the refuse to storage areas or onto the conveyors for processing. Both live bottom pit systems and the use of a bi-level receiving floor in conjunction with front-end loaders have been considered for the proposed plant. It was decided that the latter would be the most effective because of the numerous operating problems reported for live bottom pit systems. From the receiving area, the raw refuse would be transported into the processing area, as showin in Figure 5-2 and described in Section 5.2 of the flow plan.

5.6 SITE PLAN DEVELOPMENT

Utilizing the information compiled during the initial phase of this project, a comprehensive site plan for the resource recovery facility was developed. It was determined that the facility would require a refuse receiving building, a refuse processing building, a maintenance and employee building and an administrative building. To house all the required facilities, architectural plans were developed for each of these buildings. In addition to preparing the architectural designs, the structural requirements, heating, cooling, ventilation, plumbing and electrical needs for each of the facilities were determined. Further, specific installa-

126 DESIGN PRINCIPLES

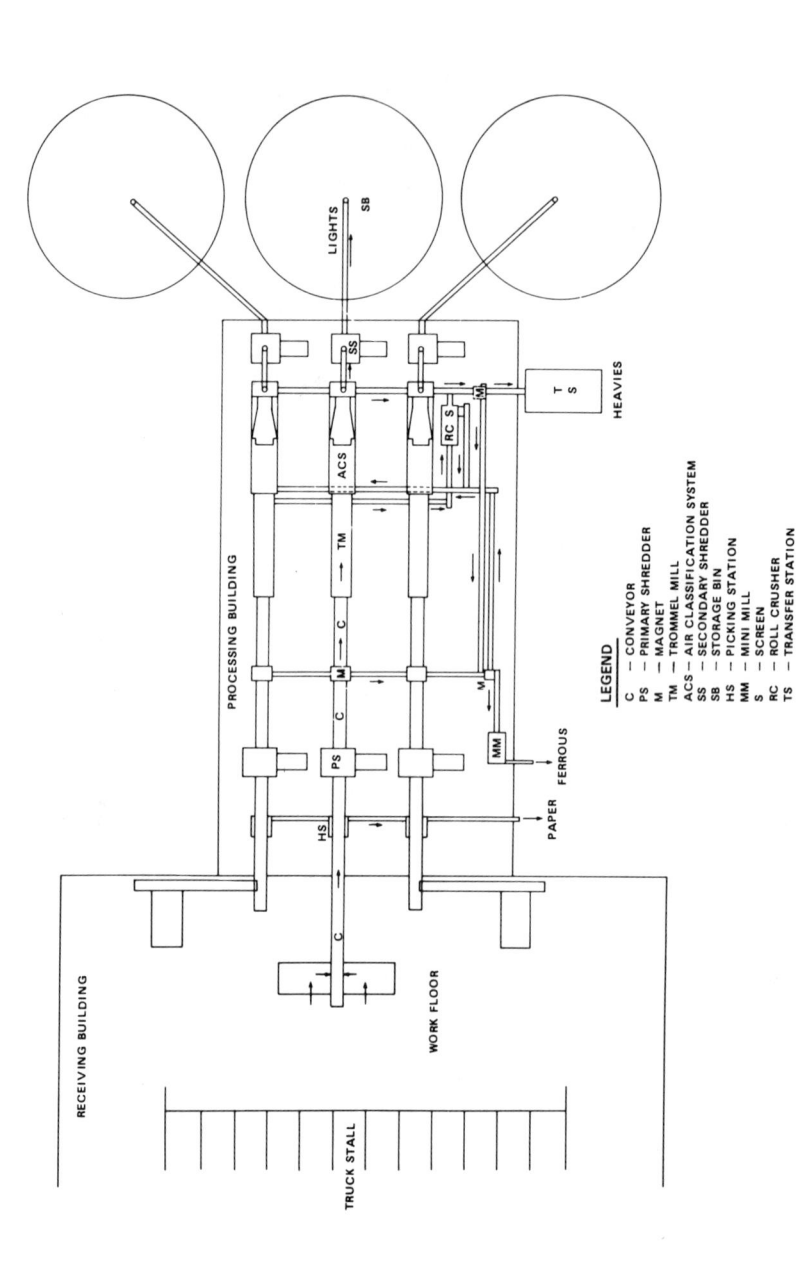

Figure 5-2. Arrangement for plant layout for resource recovery system.

PLANT DESIGN EXERCISE 127

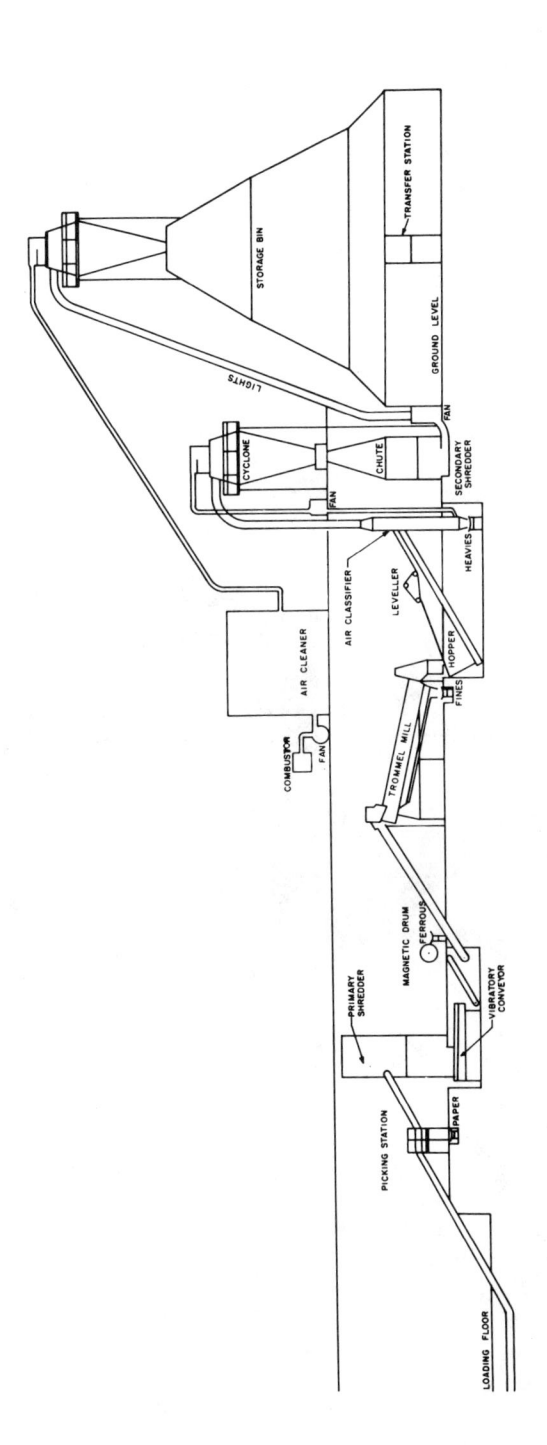

Figure 5-3. Elevation cross section of resource recovery processing line.

128 DESIGN PRINCIPLES

tion requirements for all the equipment (foundations, structural steel, etc.), safety, environmental health requirements for the facility were determined in this phase of the program. The site plan developed for the facility is presented in Figure 5-4. It was determined that the site would require a 17-acre area for the proposed resource recovery facility. The floor plans developed for the major building and the layout developed for the processing equipment for the resource recovery facility are presented

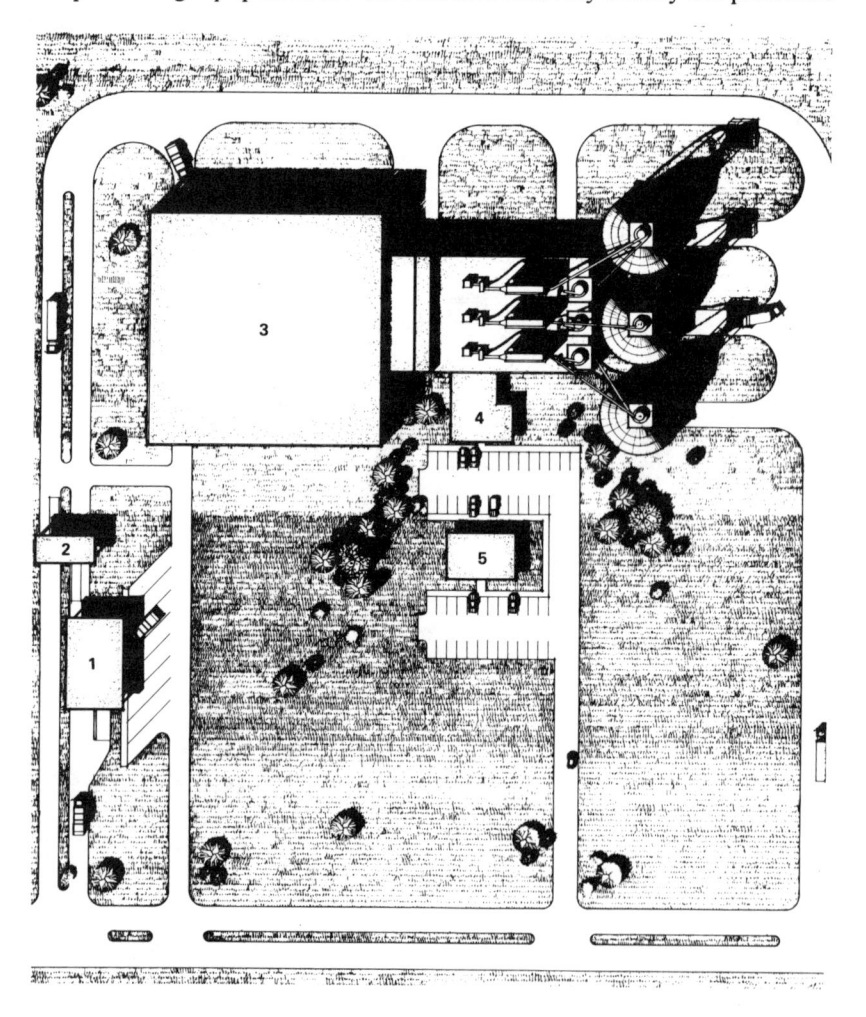

LEGEND:
1. TRUCKERS REST AREA
2. WEIGH STATION
3. PROCESSING PLANT
4. MAINTENANCE BLDG.
5. OFFICE BLDG.

Figure 5-4. Site plan.

in Figure 5-5. An elevation drawing showing the crosssection through the center of the refuse receiving and processing building is presented in Figure 5-6. An artist's rendering of the proposed facility is presented in Figure 5-7.

The truck receiving area size was determined based on the maneuvering requirements of the refuse trucks. The refuse receiving area was determined by allowing both ample maneuvering area for the front-end loaders and also a refuse storage area large enough to accommodate one day of refuse in case of a plant shutdown. The remaining area of the plant was designed to fit around the processing equipment. The processing building is constructed of concrete and glass panels on an exposed steel structural system. Concrete was chosen for its durability and chemical resistance, glass panels to allow natural light into the plant and steel because of the long spans involved.

The maintenance building provides space for repairing the processing equipment, a locker room, restrooms, a lunch room and offices for the foremen. The parking lot between the maintenance building and the office building provides for employee parking and, at the same time, serves to separate the office building from the noise of the plant. The office building provides space for payroll and billing and supervisory personnel to allow the plant to be self-sufficient. Also included is a quality control lab and engineers' office, a lunch room and conference room. A parking lot toward the main road is provided for visitors.

5.7 UPDATE

Since the AMAX study was completed, several resource recovery plants have been built and put into operation. Also, there have been new developments reported from recent laboratory and pilot studies and new equipment has been developed for processing refuse. As a result of the experiences gained from these new resource recovery facilities and with the knowledge of new equipment and new concepts, it is likely that the design developed in the AMAX study would be modified. As a result of the difficulties that have been encountered in the startup and early operation of several of the resource recovery facilities, it is also apparent that the initiation of a resource recovery plant should be approached with a good deal of caution. Detailed consideration should be given to the economic realities of the resource recovery process, as well as to the total environmental impact of these systems.

A first step in refuse processing has usually been size reduction by some type of shredding. However, recent studies have indicated that trommel-

130 DESIGN PRINCIPLES

RESOURCE RECOVERY PLANT

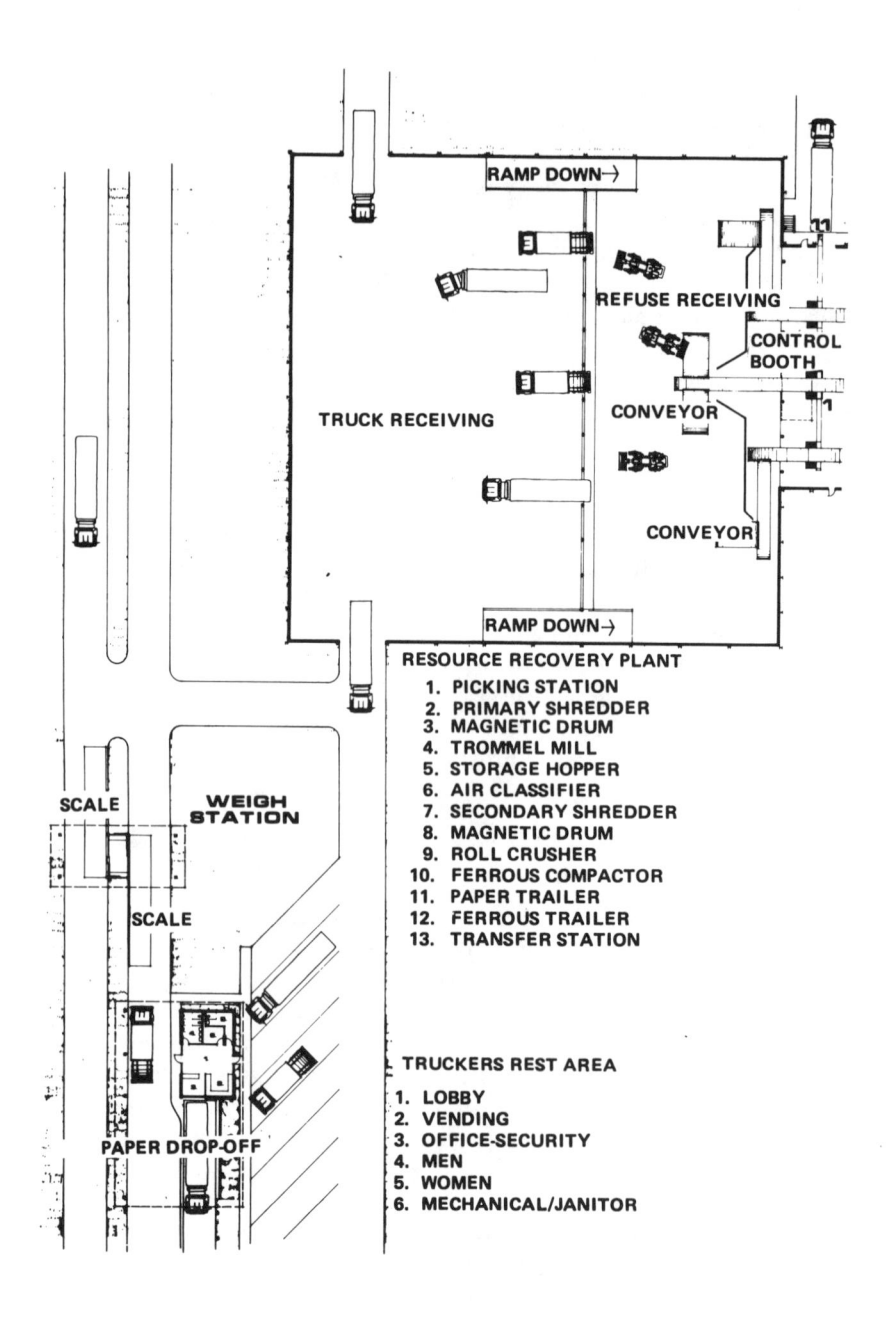

1. PICKING STATION
2. PRIMARY SHREDDER
3. MAGNETIC DRUM
4. TROMMEL MILL
5. STORAGE HOPPER
6. AIR CLASSIFIER
7. SECONDARY SHREDDER
8. MAGNETIC DRUM
9. ROLL CRUSHER
10. FERROUS COMPACTOR
11. PAPER TRAILER
12. FERROUS TRAILER
13. TRANSFER STATION

TRUCKERS REST AREA

1. LOBBY
2. VENDING
3. OFFICE-SECURITY
4. MEN
5. WOMEN
6. MECHANICAL/JANITOR

Figure 5-5. Floor plan and site arrangement.

PLANT DESIGN EXERCISE 131

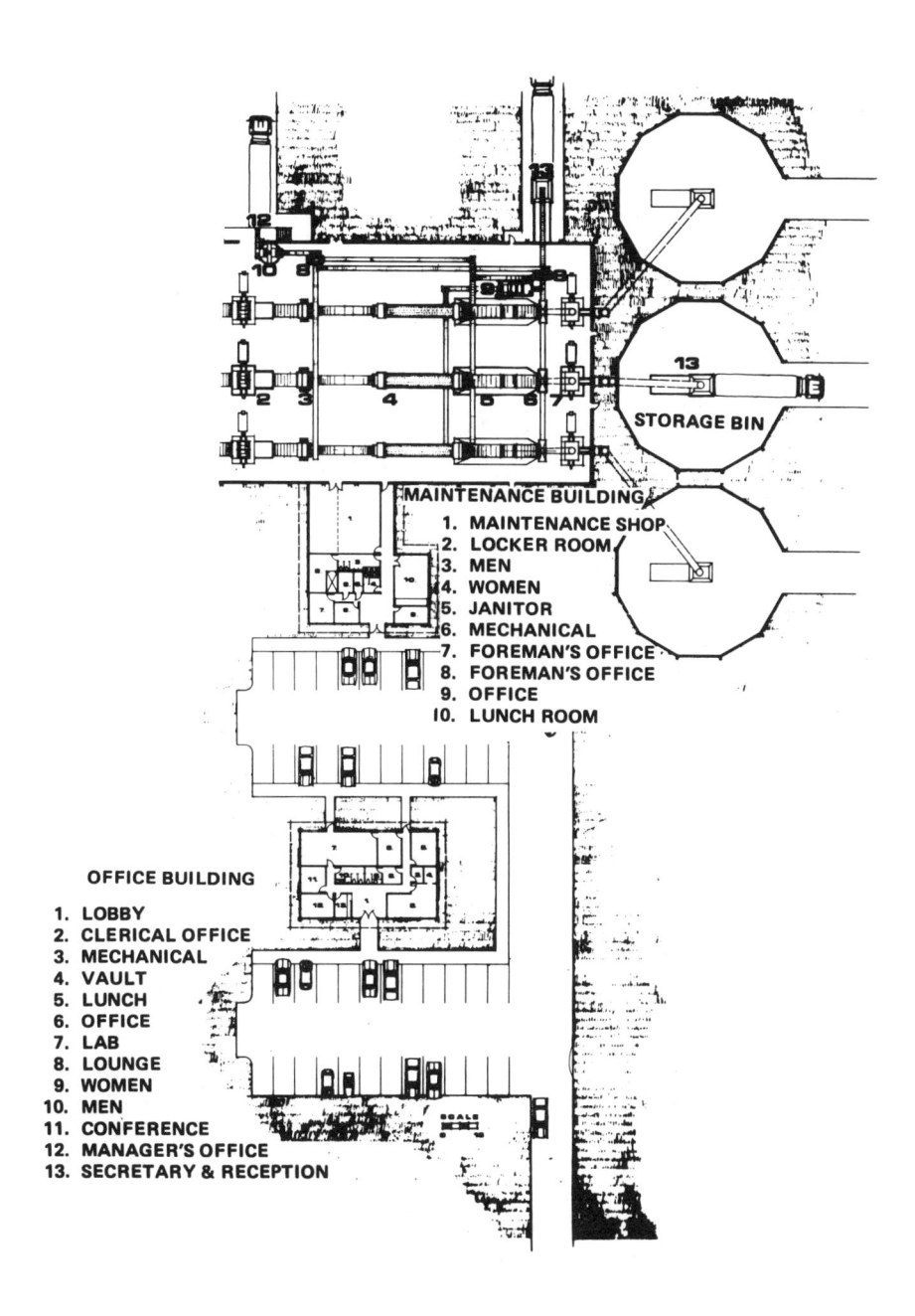

MAINTENANCE BUILDING
1. MAINTENANCE SHOP
2. LOCKER ROOM
3. MEN
4. WOMEN
5. JANITOR
6. MECHANICAL
7. FOREMAN'S OFFICE
8. FOREMAN'S OFFICE
9. OFFICE
10. LUNCH ROOM

STORAGE BIN

OFFICE BUILDING
1. LOBBY
2. CLERICAL OFFICE
3. MECHANICAL
4. VAULT
5. LUNCH
6. OFFICE
7. LAB
8. LOUNGE
9. WOMEN
10. MEN
11. CONFERENCE
12. MANAGER'S OFFICE
13. SECRETARY & RECEPTION

132 DESIGN PRINCIPLES

Figure 5-6. Cross-sectional elevation for resource recovery plant.

ing might be a more desirable first stage. Experimental results showed that about half the refuse passed through a 4-in. opening screen and contained 64% of the ferrous metal, 75% of the aluminum and 96% of the glass. The trommel's tumbling actions opened the refuse bags, allowing the bottles, cans and small trash to be sorted out. By removing this minus 4-in. material prior to shredding, the load requirements for the shredders can be significantly reduced and it may be possible to use smaller units. As the minus 4-in. material contains most of the stones, broken ceramics, glass and metal, the wear on the shredders should be reduced significantly. As most of the friable materials are removed prior to shredding, the potential for recovery is improved, particularly for the glass fraction. Similarly, ferrous metal and aluminum recovery also should be facilitated.

The many reports of explosions that have occurred in the shredding facilities reinforces the recommendation made in the AMAX study for the use of explosion suppression equipment. The poor performance reported with the use of live bottom bins for receiving the refuse supports the recommendation for a bi-level receiving floor with front-end loaders to move the refuse onto conveyors. In the AMAX study, drum magnets were recommended for ferrous separation; however, with the announced availability of stainless steel belts for the self-cleaning overhead belt magnets, it would appear desirable to replace the drum magnets for primary ferrous recovery. For the air classification stage of the process, a zig-zag unit was recommended for the proposed plant. The reports to date on the performance of the various air classification systems have been somewhat disappointing. It appears that none of the commercially available units have yet provided the level of performance required for effective air classification of the shredded refuse. In addition to the need for a unit that more effectively separates the organic fraction, there also is a need for systems better able to cope with the corrosive and erosive nature of the shredded refuse. Effective systems for dust and odor control are also needed.

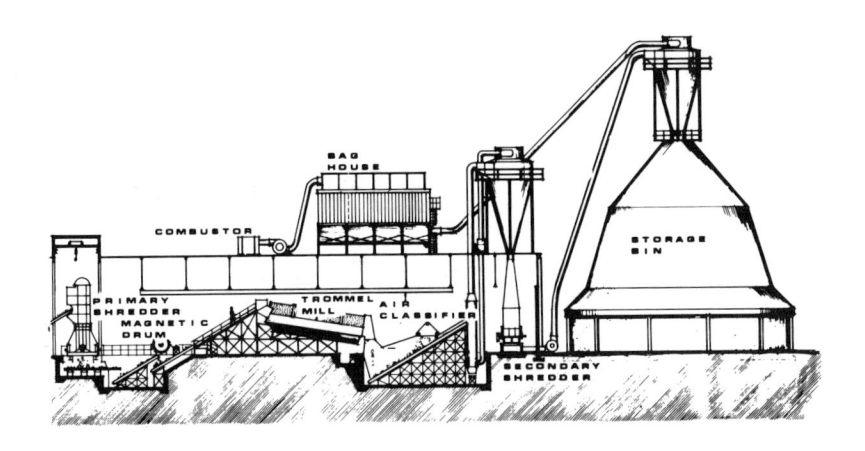

Another problem encountered in a number of the resource recovery facilities has been the storage of the shredded light fraction. The facilities developed to date have not been effective, and a better means of storing the shredded refuse is required. One concept proposed is a system that keeps the shredded refuse in continual motion. This would require an effective circulatory system maintaining the refuse in motion until it is tapped off into the fuel feeding system. Another concept is the use of small-sized (1 or 2 tons) containers, which could be stored on racks and easily moved by a forklift truck. For this system the refuse processing plant would be sized to coincide with the boiler feed schedule. On those occasions when the boiler is down for repair or modification, the shredded refuse would be conveyed to the small dump bins which, as they are filled, are moved onto a storage rack and put back into the system as required. Although the air-classified fraction has been employed success-

134 DESIGN PRINCIPLES

Figure 5-7. Proposed resource recovery plant.

fully as a fuel, a number of problems have been associated with its use. Moisture content fluctuates, greatly affecting the Btu content and the processability of the material. Yard waste and its seasonal fluctuation is also a problem. Compared with other fuels, the refuse is more difficult to transfer, store and process and has a lower energy density. Use of the refuse-derived fuel also results in the generation of considerably more ash and the potential for increased corrosion. A number of processes are being considered for producing an upgraded, refuse-derived fuel for use in boilers and industrial furnaces. Densification into pellets and drying of the light fraction are two of the processes being studied extensively. Several chemical processes also are being developed to embrittle the shredded refuse and convert it to a powder more consistent with the conventional fuels used in suspension-fired boilers. Adoption of many of these processes would result in a considerably altered process flow plan for a modern resource recovery facility.

5.8 SUMMARY

In this chapter, a 1974 University of Dayton project for AMAX Inc. was used as an example to demonstrate the sequence of steps followed in an actual design of a resource recovery plant. It is hoped that the information presented in this chapter will provide the reader with greater insight into the procedures and details required in a design project. The procedures for developing a flow plan, material balance, plant layout and site plan have been presented, along with an update to the concepts employed in 1974. This update describes more recent philosophy in the design of resource recovery systems.

The floor plan developed represents the culmination of data analysis procedures to establish market requirements, waste stream availability and the available unit operations needed to recover the required products. The material balance traces the likely flow of the waste constituents through the proposed flow plan. This material balance serves as the first step in the equipment selection process, which, in turn, is the first step in the plant layout process. The material balance identifies equipment requirements and defines the parameters for the plant layout. For the 1974 AMAX plant design project, the plant layout developed tried to use the available space in a manner as efficient as possible consistent with "state-of-the-art" to provide RDF to meet market demands and waste availability for the model community. The plant layout served as the focal point around which the site plan was developed and the building requirements established. Although the technology for resource recovery has changed considerably between 1974 and 1982 and a good deal of information has been accumulated from the many demonstration facilities, the design procedures presented are still valid for design projects in the 1980s.

CHAPTER 6

ECONOMIC ANALYSIS

6.1 INTRODUCTION

The final phase in the design of a resource recovery facility is the preparation of an economic analysis. This analysis will usually include an estimate of the capital costs for the proposed facility and the anticipated operating and maintenance (O & M) costs. A format that can be followed for this type of economic analysis has been published by EPA [86]. To better demonstrate the procedures followed for an economic analysis of a resource recovery plant, the analysis developed for the AMAX, Inc. study [85] is presented in this chapter.

6.2 ECONOMIC ANALYSIS OF PROPOSED RESOURCE RECOVERY FACILITY FOR AMAX, INC.

An economic analysis was developed for the resource recovery facility designed for AMAX, Inc. and based on the plant design exercise presented in Chapter 5. All calculations were developed for a resource recovery facility processing 350,000 tons of municipal solid refuse a year for a community located in the east, northcentral United States. All estimates were based on the economic conditions for July 1, 1974. The operating conditions assumed for the facility have been presented in Chapter 5. For a facility processing 350,000 tons of refuse a year, it is assumed that 245,900 tons of fuel, 30,600 tons of ferrous metal, 7,800 tons of newsprint and 65,700 tons of residue for landfill would be produced annually.

6.2.1 Estimated Turnkey Capital Cost

A detailed capital cost estimate was prepared for the proposed resource recovery plant. A summary of the turnkey capital costs is presented in Table 6-1 and the estimated capital cost is presented in Table 6-2. The "quantity take-off method" was used for preparing the detailed estimate. This procedure has been found by the Veterans Administration [87] to be the most accurate estimating method. Other methods, such as modulator take-off, square foot, cubic foot, unit of enclosure, unit of use, systems unit or finish unit pricing could not be used because of the varibles of usage and units involved in this study. The total lack of historical data for resource recovery plants further complicated the pricing procedure.

Table 6-1. Turnkey Capital Costs[a,b]

Buildings, Site Preparation and Equipment (see Table 6-2)	$13,100,100
Additions	
Startup costs—4 months payroll	300,000
Working capital—3 months payroll	225,000
Total Initial Investment	$13,625,000

[a]The capital cost does not include land acquisition; however, the cost to acquire 17 acres to site the plant would not exceed $170,000 (a landfill of 100 acres to accompany the resource recovery plant would add about $500,000).

[b]Financing and legal costs are not listed; however, they could be part of the total turnkey capital costs. EPA SW-157-60 guidelines recommend a value of 2% of capital cost for these items.

The turnkey capital cost estimate was based on July 1974 prices. The inflationary condition at that time created a situation in which the cost of construction materials was changing almost daily. All cost data used for the plant equipment were obtained from manufacturers' quotes; however, these quotes were based on immediate order situations. The cost summary presented in Table 6-1 is based on a detailed cost analysis of every item expected in the facility. An example of the depth of detail required can be gained from the worksheet used in the project (Table 6-3).

6.2.2 Operating Costs

The operating costs needed to carry out the required daily operations of a resource recovery facility are determined from the manpower costs, maintenance, other plant operating costs, and general and administrative costs. A summary of these operating costs is presented in Table 6-4 and

Table 6-2. Summary of Estimated Capital Costs

Description		Material	Labor	Subcontract	Total
Plant entrance and grounds		$154,700	$ 85,290	$ 194,640	$ 434,630
Truck crew and rest area		3,920	4,030	59,375	67,325
Paper dropoff		4,110	4,375	3,675	12,160
Scale house		1,132	1,371	16,905	19,408
Scales and pits		7,267	5,608	40,400	53,275
Industrial building		189,175	196,180	1,407,660	1,793,015
Refuse receiving area		110,465	65,415	425,550	601,430
Electrical substation and motor controls				890,000	890,000
Paper picking station		1,480	1,890	5,720	9,090
Primary shredding		465,475	41,430	150,010	656,915
Primary magnetic separation		75,610	3,505	33,300	112,415
Fines separation		182,370	15,150	25,360	222,880
Fines processing		54,570	5,770	39,500	99,840
Air classification		529,235	67,645	145,580	742,460
Heavies processing		15	60	47,200	47,275
Final ferrous processing		98,300	13,650	27,645	139,595
Secondary shredding		745,760	53,600	20,900	820,260
Lights storage		713,095	112,255	1,085,350	1,910,700
Miscellaneous conveyors		1,075	1,405	129,525	132,005
Office building		5,675	4,975	110,265	120,915
Maintenance building		6,505	6,705	129,395	142,605
Sprinkler protection				154,000	154,000
Explosion protection				66,000	66,000
Carbon dioxide system				55,000	55,000
Control room protection				3,500	3,500
		3,349,934	690,309	5,266,455	9,306,698
Sales tax	4-½%	150,747			150,747
Insurance	22%		151,868		151,868
		3,500,681	842,177	5,266,455	9,609,313
Field offices					83,000
Field supervision					185,000
Main office expense					710,000
Small tools					46,000
Profit					600,000
					11,233,313
Contingency	12-½%				1,441,687
Engineering					425,000
TOTAL COSTS					13,100,000

140 DESIGN PRINCIPLES

Table 6-3. Example of Detailed Breakdown Required for Estimated Capital Costs

Description	Quantity	Material	Labor	Subcontract	Total
Secondary Shredding					
Transport tubes	6 tons			5,800	5,800
Shredders	3 pieces	367,500	11,700	9,500	388,700
Grout	14 ft^3	340	420		760
Anchor bolts	60 pieces	120	480		600
Concrete foundation	274 yd^3	17,800	27,400		45,200
Air sweep exit pans, etc.	3 pieces	360,000	13,600	5,600	379,000
Painting	Linear surface			X	X
TOTAL COSTS		745,760	53,600	20,900	820,260
Lights Storage					
Storage bins	3 pieces	552,000		828,000	1,380,000
Concrete footings	432 yd^3	11,230	2,600		13,830
Concrete columns	48 yd^3	1,250	720		1,970
Concrete slab structural	1,810 yd^3	47,060	18,100		65,160
Concrete on ground	300 yd^3	7,800	1,800		9,600
Form footing	3,890 ft^2	1,360	2,530		3,890
Form columns	3,460 ft^2	1,730	6,575		8,305
Form slab	15,550 ft^2	9,640	22,550		32,190
Form trenches	5,040 ft^2	2,520	9,800		12,320
Reinforcing steel	160 tons	72,000	33,600		105,600
Subgrade	12,210 ft^2	----	3,050		3,050
Trowel and cure	24,420 ft^2	245	3,900		4,145
Meshand vapor barrier	13,430 ft^2	1,880	670		2,550
Anchor bolts	170 pieces	280	1,360		1,640
Excavation and backfill	1,200 yd^3			7,200	7,200
Edge angles	1,260 ft	4,100	5,000		9,100
Grizzly bars	5,040 ft^2			22,000	22,000
Steel siding	10,850 ft^2			41,000	41,000

detailed in Tables 6-5 to 6-10. The manpower costs were developed from the personnel requirements established and the estimated wages and overhead for each position. As stated, the plant was designed to process refuse for two 8 hr/day shifts, 5 day/wk. The third shift of the work week was to be used for maintenance and housecleaning. During the weekends, only the refuse receiving and equipment maintenance activities were to be in operation. For most processing shifts only two lines would be operated. The third processing line would be required about 5% of the time to handle excess refuse received during the peak seasons and probably an additional 5% to compensate for processing lines that had been shut down. The personnel requirements developed for the proposed plant were based on two processing lines operating two shifts; however, sufficient flexibility was provided to cover the normal amount of employee absenteeism and vacations (see Table 6-5). The work force required

Table 6-4. Operating Costs Using 1974 Rates (excludes interest on investment)

1. Manpower Costs (Tables 6-5 and 6-8)		$ 879,682
2. Maintenance Material (Tables 6-9 and 6-10)		681,741
3. Other Plant Costs[a] (Table 6-11)		681,741
4. General and Administrative Costs (Table 6-12)		41,960
	SUBTOTAL	$2,324,003
5. Depreciation (Table 6-10)		1,423,865
6. Startup Costs		15,000
	TOTAL	13,762,868
Operating cost per ton $\dfrac{\$2,324,003}{350,000}$		$ 6.64
Total cost per ton[b] $\dfrac{\$3,762,868}{350,000}$		$ 10.75

[a]Residue disposal was not included in these calculations. At an estimated disposal cost of $4.50/ton, the cost for 65,700 ton would be $295,650, or an additional $0.85/ton.
[b]Annual interest costs have not been included but should be recognized as an operating cost.

includes an administrative staff of six during the first eight-hour shift and a crew of 21 to operate the refuse processing system. Based on the staff requirements established (see Table 6-6), prevailing wage rates and determined overhead, total manpower costs were determined (see Table 6-7). An example of the line item manpower costs used is presented in Table 6-8.

Due to the limited experience of resource recovery operations, it was difficult to estimate maintenance materials costs and, in most cases, general engineering guidelines were used. However, because of the extensive maintenance requirements reported for shredding operations, it was decided to consider shredder maintenance separately. Total estimated maintenance material costs are presented in Table 6-9. The data used for calculating the shredder costs were obtained from manufacturers' data and shredding operation reports. Of the total installed cost of equipment, including foundation, 5% was used as the basis for calculating the yearly maintenance material and outside maintenance service cost. The shredding equipment also was included in this calculation to cover any unforeseen maintenance requirements that might be needed. Although the buildings were considered to have been designed for maintenance-free operation, 1% of the installed costs were allowed for calculating building material needs. The maintenance material costs for all equipment were compiled and are presented in Table 6-10. (This table also includes depreciation costs for the facility.)

In addition to manpower and maintenance costs, a number of other

Table 6-5. Proposed Work Schedule

Day	Monday			Tuesday			Wednesday			Thursday			Friday			Saturday			Sunday		
Shift	1st	2nd	3rd	1st	2nd	3rd	1st	2nd	3rd	1st	2nd	3rd	1st	2nd	3rd	1st	2nd	3rd	1st	2nd	3rd
Weighmaster	1	1	1	1	1	1	1	1	1	1	1	1	1	1	1	1	1	1			
Front-end loader operator	2	2		2	2		2	2		2	2		2	2							
Control tower operator	2	2		2	2		2	2		2	2		2	2							
Process operator	2	2		2	2		2	2		2	2		2	2							
Assistant process operator	1	1		1	1		1	1		1	1		1	1							
Product equipment operator	2	2		2	2		2	2		2	2		2	2							
Maintenance man	2	2	3	2	2	3	2	2	3	2	2	3	2	2	3	1	1	1	1	1	1
Maintenance helper	1	1	1	1	1	1	1	1	1	1	1	1	1	1	1	1	1	1	1	1	1
Utility man I	1	1		1	1		1	1		1	1		1	1							
Utility man II	1			1			1			1			1								
Receiving area attendant	1			1			1			1			1								
Janitor		1			1			1			1			1							
Handpicker	4	4		4	4		4	4		4	4		4	4							
Foreman	1	1		1	1		1	1		1	1		1	1							
Plant manager	1			1			1			1			1								
Assistant plant manager	1			1			1			1			1								
Billing clerk	1			1			1			1			1								
Accounting clerk	1			1			1			1			1								
Secretary	1			1			1			1			1								
Quality control technician	1			1			1			1			1								
TOTAL	27	20	8	27	20	8	27	20	8	27	20	8	27	20	8	5	5	5	5	5	5

Table 6-6. Required Manpower

I. Processing Staff
   1. Weighmaster (required 24 hr/day, 7 days/wk)
   2. Front-end loader operators (2/shift, 1st and 2nd shifts)
   3. Control tower operators (2/shift, 1st and 2nd shifts)
   4. Process operators (2/shifts, 1st and 2nd shifts)
   5. Assistant process operator (1/shift, 1st and 2nd shifts)
   6. Product equipment operators (2/shift, 1st and 2nd shifts)
   7. Maintenance men (2/shift, 1st and 2nd shifts; 3 on 3rd shift; 1 on each weekend shift)
   8. Maintenance helper (1 on all shifts, 24 hr/day, 7 day/wk)
   9. Utility man I (1 on all shifts, 24 hr/day, 7 day/wk)
   10. Utility man II 1/shift, 1st and 2nd shifts)
   11. Receiving area attendant (1 on 1st shift only)
   12. Janitor (1 on 3rd shift only)
   13. Handpickers (4/shift, 1st and 2nd shift)
   14. Foreman (required 24 hr/day—7 day/wk)
II. Administrative Staff (1st shift only)
   15. Plant manager
   16. Secretary
   17. Assistant plant manager
   18. Billing clerk
   19. Accounting clerk
   20. Quality control technician

operating costs need to be considered. These include operation maintenance for the front-end loaders, outside security force, taxes, insurance, water, sewer and power costs, as well as other miscellaneous items. The cost calculated for these operations is presented in Table 6-11. It should be noted that the cost for residue disposal was not included in these calculations. Based on 1974 data, it was determined that landfill disposal at $4.50/ton would result in an added annual operating cost of $295,650 for residue disposal.

General and administrative operating costs are detailed in Table 6-10. Depreciation costs were calculated on a straight line basis, as shown in Table 6-12. A period of 20 years was used as the accounting life for all buildings, 10 years for the shredder scale and electrical system, 3 years for the plant vehicles and 7 years for all other equipment. A startup cost of $300,000 was anticipated and, over 20 years, this would amount to $15,000/yr, as shown in Table 6-4. Table 6-4 does not include the interest to be paid on the capital expenditure for the proposed facility. This omission was at the request of the project sponsor; however, it is a cost that must be taken into account in determining total annual operating costs for a facility.

144 DESIGN PRINCIPLES

Table 6-7. Manpower Costs

Position	Annual Cost for Each Position	Number Required	Total Cost
Weighmaster	$13,410.19	4	$ 53,640.76
Front-end loader operator	14,568.23	4	58,272.92
Control tower operator	14,568.23	4	58,272.92
Process operator	13,073.95	4	52,295.80
Assistant process operator	13,073.95	2	26,147.90
Product equipment operator	13,073.95	4	52,295.80
Maintenance man	16,713.08	4	66,852.32
Maintenance man	17,935.67	4	71,742.68
Maintenance helper	15,681.33	4	62,725.32
Utility man I	15,681.33	4	62,725.32
Utility man II	13,073.95	2	26,147.90
Receiving area attendant	10,871.86	1	10,871.86
Janitor	12,471.00	1	12,471.00
Handpicker	10,871.86	8	86,974.88
Foreman	20,529.89	4	82,119.56
Plant manager	28,366.39	1	28,366.39
Assistant plant manager	25,007.89	1	25,007.89
Billing clerk	9,627.60	1	9,627.60
Accounting clerk	11,041.20	1	11,041.20
Secretary	9,627.60	1	9,627.60
Quality control technician	12,454.80	1	12,454.80
TOTAL		60	$879,682.42

Table 6-8. Example of Line Item Manpower Costs

Weighmaster, Class II

Annual wage, $4.32 × 2080 hr	$8,985.60	
Planned overtime, $4.32 × 12.5 × 8 hr × 150%	648.00	
Holiday premium, $4.32 × 6 × 8 hr	207.36	
Overtime, $4.32 × 2080 × 5% × 150%	673.92	$10,514.88
Fringe Costs		
Unemployment compensation	$ 126.00	
Workmen's compensation	415.34	
Insurance package	1,020.00	
FICA (5.85%)	615.12	
Retirement	718.85	2,895.31
TOTAL		$13,410.19

Table 6-9. Total Maintenance Material and Outside Maintenance Cost

Primary shredders		$116,730
Secondary shredders		73,538
All other equipment and buildings		491,473
(Table 6-10)	TOTAL	$681,741

6.3 SUMMARY

In this chapter, the factors used in determining the capital and operating costs for a resource recovery system were presented through the economic analysis developed for the resource recovery facility designed in Chapter 5. As shown, the capital cost is a systematic compilation of the costs for all the equipment required, for the installation of that equipment, and for the facility needed to house the equipment and operations for the plant. In addition, the capital cost includes the engineering, legal and startup costs needed to allow the facility to begin operations. The annual operating cost, frequently referred to as the O&M cost, represents the total manpower, maintenance, general and administrative costs, as well as other plant operating costs such as security, taxes, insurance, water/sewer, power, etc. The annual operating cost divided by the total number of tons processed each year represents the operating cost per ton processed. Adding to this number the depreciation and amortization cost on a per ton basis gives the total cost per ton. This cost provides a basis for determining the revenues required per ton to successfully operate the resource recovery facility.

146 DESIGN PRINCIPLES

Table 6-10. Capital Cost, Depreciation and Maintenance Materials

Item	Contractor's Direct Charge	Delivered Invoice Cost	Installation Cost	Foundations	Sales Tax
Sitework	$ 372,630				$ 6,962
Driver's lounge building	66,200				176
Paper trailer dock	12,160				185
Weigh station building	19,408				51
Weigh station pit	14,275				327
Industrial area	1,793,015				6,263
Receiving and storage area	48,430				651
Picking platforms	9,090				67
Office building	108,425				255
Maintenance building	93,815				293
Weigh scales		$ 39,000			
Front-end loaders		96,000			4,320
Primary shredders		447,000	$ 22,200	$ 82,715	20,946
Magnetic separation		75,500	3,400	3,515	3,402
Trommel mills		175,550	7,700	19,130	8,207
Roll crusher and vibrating screen		52,260	4,000	13,840	2,456
Air classification		525,000	79,000	138,460	23,816
Transfer station—heavies		25,000		675	
Minimill		92,000	6,000	17,195	4,424
Secondary shredder		367,500	21,200	52,360	17,359
Air sweep exit pans		360,000	19,200		16,200
Light storage system		552,000	828,000	395,700	32,089
Transfer system		135,000			
Electrical substation and motor control		890,000			
Sprinkler protection		154,000			
Explosion protection		66,000			
CO_2 system		55,000			
Control room protection		3,500			
TV security system		35,000			
Power sweeper		18,000			
Forklift truck		9,000			
Conveyors		666,900		4,380	2,298
Conveyor covers		93,825			
Desks, chairs, etc.		22,245			
Machine shop equipment		45,000			
Engineering		475,000			
TOTAL COSTS	$2,537,448	$5,475,280	$990,700	$727,970	$150,747

Contractor's Overhead & Contingency	Total Cost	Accounting Life (yr)	Annual Depreciation	Estimated Maintenance (%)	Annual Maintenance Material Cost
$ 131,346	$ 510,938	20	$ 25,547	1	$ 5,109
23,355	16,600	20	830	1	166
6,812	26,271	20	1,314	1	263
5,020	19,622	20	981	1	196
617,184	2,416,462	20	120,823	1	24,165
15,780	64,861	20	3,243	1	649
3,200	12,357	20	618	1	124
38.059	146,739	20	7,337	1	1,467
32,968	127,076	20	6,354	1	1,271
13,670	52,670	10	5,267	5	2,634
22,450	122,770	3	40,923	5	4,467
194,736	767,597	10	76,760	5	38,380
28,876	114,693	7	16,385	5	5,735
70,966	281,553	7	40,222	5	14,078
24,516	97,072	7	13,867	5	4,854
250,245	1,016,521	7	145,217	5	50,826
9,019	34,694	7	4,956	5	1,735
40,459	160,078	7	22,297	5	8,004
157,397	615,816	10	61,582	5	30,791
123,324	518,724	7	74,103	5	25,936
617,601	2,425,390	7	346,484	5	121,270
47,300	182,300	7	26,043	5	9,115
311,067	1,201,067	10	120,107	5	60,053
54,120	208,120	7	29,731	5	10,406
23,280	89,280	7	12,754	5	4,464
19,300	74,300	7	10,614	5	3,715
1,240	4,740	7	10,614	5	3,715
12,300	47,300	7	6,757	5	2,365
6,305	24,305	3	8,102	5	1,215
3,180	12,180	3	4,060	5	609
234,327	907,905	7	129,701	5	45,395
32,971	126,796	7	18,114	5	6,340
7,787	30,032	7	4,290	5	1,502
15,800	60,800	7	8,686	5	3.040
17,640	492,640	20	24,632	—	-0-
$3,217,855	$13,100,000		$1,423,865		$491,473

Table 6-11. Other Plant Operating Costs

1. Front-end loader operation and maintenance	$ 48,384.00
2. Outside security force	30,028.80
3. Personal property taxes	177,908.00
4. Real estate taxes	120,435.000
5. Insurance	65,500.00
6. Water and sewer	1,440.00
7. Miscellaneous	18,000.00
8. Power costs	258,924.000
	$720,619.80

Table 6-12. Annual General and Administrative Costs

1. Postage and office supplies	$1,200
2. Janitorial service (outside contractor)	1,680
3. Telephone	2,880
4. Monthly accounting service	3,600
5. Year end audit	3,000
6. Lab supplies	2,400
7. Landscaping (outside contractor)	3,000
8. Travel and entertainment	6,200
9. Contributions, subscriptions and miscellaneous	6,000
10. Employee services	6,000
11. Public relations	6,000
TOTAL	$41,960

EPILOGUE

The manuscript for this book was written and assembled over a period of nearly three years, during which time the technology for solid waste disposal has undergone considerable changes. Most noticeable is the emergence, acceptance and continuing growth of resource recovery technology for the management of municipal solid waste. Even during this current period of tight money and reduced federal support, the number of communities considering and instituting resource recovery systems continues to grow. The more stringent environmental codes, increasing costs of sanitary landfill and incineration, and the rising costs of energy will continue to serve as stimulants for instituting resource recovery systems. This situation is further enhanced as more plants come on line and develop successful records of operation. However, the industry is still confronted with problems common to many emerging technologies. Some plants have been forced to shut down either temporarily or permanently, while others continue to be plagued by operating problems. This situation is exacerbated, of course, by the basic nature of municipal solid waste and the fact that its composition and characteristics are constantly changing and are not easily processed on conventional equipment designed for materials with more uniform characteristics.

As discussed in Chapter 1, a number of different resource recovery systems are available to a community for managing its wastes. And, as would be anticipated, some are more proven than others. In addition, there is a diversity of unit operations available that can be coupled together to process municipal solid wastes to recover useful products. Unfortunately, design and operating performance data are still very much fragmented. However, during the past three years there has been a significant increase in the data describing performance and operating behavior of a variety of resource recovery systems and plant operations [88–92]. During these past three years, we also have developed a greater understanding of the basic principles for a wide range of waste processing systems. Much of our understanding comes from the recent data published about the operation of the first generation resource recovery systems and pilot programs. In particular, we have obtained better insight into shredding, air classification, screening, magnetic separation, and material handling (transfer and storage) operations. In addition, considerably more has been learned about the combustion of raw and processed refuse.

Although mass burning systems are the most commercially proven, the production and combustion of RDF is receiving much greater acceptance. In addition, the technologies for glass, metal and plastics recovery, as well as composite production, are demonstrating substantial advancement [93-95]. During the past three years, acceptance of resource recovery as a viable technology for solid waste management has undergone many trials and tribulations and has been the subject of considerable public scrutiny. However, it is fair to say that resource recovery is being accepted and embraced by many communities around the country. Based on a March, 1982 newsletter from the U.S. Conference of Mayors [96], there are 83 energy and/or material recovery facilities in the U.S. and Canada that are either in operation, shakedown construction or design, with a total design capacity of 57,000 tons per day. Of these plants, 42 are in operation or shakedown and have a design capacity of 17,000 tons per day. There are 19 under construction and 15 about to begin construction; 7 plants have suspended operations but have hopes of renewing operations at some future time. In addition to these 83 facilities, some 52 communities have initiated a variety of planning programs for instituting resource recovery facilities. If all these plans come to fruition, there will be a resource recovery design capacity of approximately 90,000 tons per day by 1987.

This acceptance and growth of resource recovery systems for community solid waste management programs extends the need for effective plant design procedures. The 20 steps developed by Apple [83] (see Chapter 4) can serve as an effective basic guideline for most resource recovery plant design programs.

REFERENCES

1. "Resource Recovery Activities," *NCRR Bull.* 11(1):15-24 (March 1981).
2. Office of Technology Assessment. "Materials and Energy from Municipal Waste," U.S. Government Printing Office, Washington, DC (1979), p. 255.
3. Arella, D. G., and Y. M. Garbe. "Mineral Recovery from the Noncombustible Fraction of Municipal Solid Waste," Office of Solid Waste Management, SW-82d.1, U.S. Environmental Protection Agency (1975).
4. Bendersky, D., et al. "St. Louis Refuse Fuel Demonstration Plant—Technical and Economic Performance," IEEE Catalog No. 75CH1008-2CRE (November 1975), pp. 7-13.
5. "New Orleans Resource Recovery Facility Implementation Study Equipment, Economics, Environment," National Center for Resource Recovery, Inc., Washington, DC (September 1977).
6. Wittman, T. J., D. J. McCabe and M. C. Eifert. "A Technical Environmental and Economic Evaluation of the Wet Processing System for the Recovery and Disposal of Municipal Solid Waste," Office of Solid Waste Management, U.S. Environmental Protection Agency (January 1975).
7. Office of Technology Assessment. "Materials and Energy from Municipal Waste," U.S. Government Printing Office, Washington, DC (1979), p. 257.
8. Wiles, C. C. "Summaries of Combustion of Refuse-Derived Fuels and Densified Fuels," in *Municipal Solid Waste: Resource Recovery*, David W. Shulz, Ed., U.S. Environmental Protection Agency, EPA-600/9-81-002C (March 1981), p. 144.
9. "Technology Review—Producing and Burning D-RDF," *NCRR Bull.* 10(4):86-90 (December 1980).
10. Hasselriis, F. "The Greater Bridgeport, Connecticut Waste-to-Power System," in *Proceedings of the 1980 National Waste Processing Conference* (New York: American Society of Mechanical Engineers, 1980), pp. 435-446.
11. Office of Technology Assessment. "Materials and Energy from Municipal Waste," U.S. Government Printing Office, Washington, DC (1979), p. 261.
12. Levy, S. J. "The Conversion of Municipal Solid Waste to a Liquid Fuel by Pyrolysis," in *Conversion of Refuse to Energy*, IEEE Catalog No. 75CH1008-2CRE (November 1975), pp. 226-231.
13. Legille, E., F. A. Berczynski and K. G. Heiss. "A Slagging Pyrolysis Solid Waste Conversion System," in *Conversion of Refuse to Energy*, IEEE Catalog No. 75CH1008-2CRE (November 1975), pp. 232-257.

14. Appell, H. R., I. Wender and R. D. Miller. "Conversion of Urban Refuse to Oil," Bureau of Mines Report 25, PB192414 (May 1970).
15. Gupta, D. V., W. L. Kranich and A. H. Weiss. "Catalytic Hydrogenation and Hydrocracking of Oxygenated Compounds to Liquid and Gaseous Fuels," *Ind. Eng. Chem. Proc. Des. Dev.* (April 15, 1976), p. 256.
16. Walter, D. K., and C. Rines, "Refuse Conversion to Methane (RefCom): A Proof-of-Concept Anaerobic Digestion Facility," in *Proceedings of the 1980 National Waste Processing Conferences* (New York: American Society of Mechanical Engineers, 1980), pp. 85-92.
17. Office of Technology Assessment. "Materials and Energy from Municipal Waste," U.S. Government Printing Office, Washington, DC (1979), p. 262.
18. James, S. C., and C. W. Rhyne. "Methane Production from the Mountain View Landfill," in *Energy from Waste*, T. C. Frankiewicz, Ed., Vol. I in the series *Design and Management for Resource Recovery*, P. A. Vesilind, Series Ed. (Ann Arbor, MI: Ann Arbor Science Publishers, Inc., 1980), pp. 21-31.
19. Sullivan, P. M., M. H. Stancyzk and M. J. Spendlove. "Resource Recovery from Raw Urban Refuse," U.S. Bureau of Mines, RI 7760 (1973).
20. Webb, M., and L. Whalley. "Household Waste Sorting System," in *Raw Materials, Vol. I*, Commission of the European Communities (January 1979), p. 56.
21. Wiley, J. S., F. E. Gartrell and H. G. Smith. "Concept and Design of the Joint U.S. Public Health Service—Tennessee Valley Authority Composting Project, Johnson City, Tennessee," *Compost Sci.*, 7(2):11-14 (Autumn 1966).
22. Niessen, W. R., and S. H. Chansky. "The Nature of Refuse," in *Proceedings of the 1972 National Incineration Conference* (New York: American Society of Mechanical Engineers, 1972).
23. Doggett, R. M., M. K. O'Farrell and A. L. Watson. "Forecasts of the Quantity and Composition of Solid Waste, U.S. Environmental Protection Agency, EPA-600/5-80-001 (1980).
24. Smith, F. A. "Quantity and Composition of Post-Consumer Solid Waste: Material Flow Estimates for 1973 and Baseline Future Projections," *Waste Age* (April 1976).
25. Wilson, D. G., Ed. *Handbook of Solid Waste Management* (New York: Van Nostrand Reinhold Company, 1977).
26. Ruf, J. "Particle Size Spectrum and Compressibility of Raw and Shredded Municipal Solid Waste," PhD Thesis, The University of Florida (1974).
27. Rogers, H. W., and S. J. Hitte. *Solid Waste Shredding and Shredder Selection*, Office of Solid Waste Management, SW-140, U.S. Environmental Protection Agency, EPA/530/SW-140 (1975).
28. Midwest Research Institute. *Study of Processing Equipment for Resource Recovery Systems, Vol. 1*, U.S. Environmental Proteciton Agency (December 1978), p. 46.
29. Alter, H., NCRR. Private communication (June 1974).
30. Midwest Research Institute. "Size-Reduction Equipment for Municipal Solid Waste," PB 226 551 (1973).
31. Midwest Research Institute. *Study of Processing Equipment for Resource Recovery Systems, Vol. III—Field Test Evaluation of Shredders*, U.S. Environmental Protection Agency (April 1979).
32. Trezek, G. J. "Significance of Size Reduction in Solid Waste Management," U.S. Environmental Protection Agency, EPA-600/2-77-131 (July 1977).

33. Drobny, N. L., H. E. Hull and R. F. Testin. *Recovery and Utilization of Municipal Solid Waste*, U.S. Environmental Protection Agency, Solid Waste Management SW-10C (1971).
34. Taggart, A. F. "Screen Sizing," in *Handbook of Mineral Dressing* (London: John Wiley & Sons, Inc., 1927), pp. 7-01–7-72.
35. Sullivan, J. F. *Screening Technology Handbook* (Dallas, TX: Triple/S Dynamics, Inc., 1975), p. 26.
36. "How to Select the Right Overstrom High-Speed Vibrating Screen," Overstrom, Dallas, TX.
37. "Rotascreen," in *Rotary Screen* (Dallas, TX: Triple/S Dynamics, Inc., 1977).
38. Fiscus, D. E., et al. "Evaluation of the Performance of the Disc Screens Installed at the City of Ames, Iowa Resource Recovery Facility," in *Proceedings of the 1980 National Waste Processing Conference* (New York: American Society of Mechanical Engineers, 1980), pp. 485-495.
39. Taggart, A. F. *Handbook of Mineral Dressing* (London: John Wiley & Sons, Inc., 1927), pp. 5-01–5-133.
40. Hill, R. M., Triple/S Dynamics. Personal communication, Dallas, TX (July 1979).
41. Bernheisel, J. F., P. M. Bagalman and W. S. Parker. "Trommel Processing of Municipal Solid Waste Prior to Shredding," in *Proceedings of the Sixth Mineral Waste Utilization Symposium*, E. Aleshin, Ed. (Chicago: IIT Research Institute, 1978), pp. 254-260.
42. Woodruff, K. L. "Preprocessing of Municipal Solid Waste for Resource Recovery with a Trommel," *Trans. Soc. Min. Eng. AIME* 260(3):201-204 (1976).
43. Woodruff, K. L., and E. P. Bales. "Preprocessing of Municipal Solid Waste for Resource Recovery with a Trommel-Update," in *Proceedings of the 1978 National Waste Processing Conference*, Chicago (May 1978).
44. Savage, G. M., L. F. Diaz and G. J. Trezek. "RDF: Quality Must Precede Quantity," *Waste Age* (April 1978).
45. Savage, G. M., and G. J. Trezek. "Screening Shredded Municipal Solid Waste," *Compost Sci.* (January/February 1976).
46. Hecht, N. L., et al. "Obtaining Improved Products from the Organic Fraction of Municipal Solid Waste," U.S. Environmental Protection Agency, EPA-600/2-80-121 (1980).
47. Midwest Research Institute. *Study of Processing Equipment for Resource Recovery Systems, Vol. II, Magnetic Separators, Air Classifier and Ambient Air Emissions Tests*, U.S. Environmental Protection Agency, 4213-D.
48. Simister, B. W., and G. Savage. "Test Report—Comparative Study of Air Classifiers, Baltimore County, Maryland," U.S. Environmental Protection Agency, Draft Copy (1979).
49. Murray, D. L. "Air Classifier Performance and Operating Principles," paper presented at 1978 National Waste Processing Conference, Chicago, May 7-10, 1978.
50. "Air Classification: Advancing the Art of Resource Recovery," special issue, *Waste Age* 8(3), 1977.
51. Nollet, A. R., and P. E. Sherwin. "New Horizons for the Solid Waste Processing and Resource Recovery Industry," AENCO, Inc., New Castle, DE.
52. Nollet, A. R., and E. T. Sherwin. "Air Classify First, Then Shred," paper

presented at the AMSE Eighth National Waste Processing Conference and Exhibit, Chicago, May 7–10, 1978.
53. Cederholm, C. "Using Air Technology to Recover Resource from Solid Waste," *ASAE Int.* (1979).
54. Worrell, W. A. "Testing and Evaluation of Three Air Classifier Throat Designs," Duke Environmental Center, Department of Civil Engineering, Duke University, Durham, NC.
55. Boettcher, R. A. *Air Classification of Solid Wastes*, Office of Solid Waste Management, SW-30C, U.S. Environmental Protection Agency (1972).
56. Midwest Research Institute. *Study of Processing Equipment for Resource Recovery Systems, Vol. 1*, U.S. Environmental Protection Agency (December 1978), p. 60.
57. Midwest Research Institute. *Study of Processing Equipment for Resource Recovery Systems, Vol. 1*, U.S. Environmental Protection Agency (December 1978), p. 61.
58. Midwest Research Institute. *Study of Processing Equipment for Resource Recovery Systems, Vol. 1*, U.S. Environmental Protection Agency (December 1978), p. 62.
59. Chesney, R. D. R., and V. R. Degner. "Hydraulic, Heavy Media and Froth Flotation Processes Applied to Recovery of Metals and Glass from Municipal Solid Waste Streams," paper No. 38B, presented at the AIChE 78th National Meeting, August 18–21, 1974.
60. Michaels, E. L. "Heavy Media Separation of Aluminum from Municipal Solid Waste," paper presented at AIME 103rd Annual Meeting, Dallas, TX, February 1974.
61. Booker, M. "Inorganic Resource Recovery and Solid Fuel Preparation from Municipal Trash," in *Proceedings of the Fourth Mineral Waste Utilization Symposium* (Chicago: IIT Research Institute, 1974), pp. 86–94.
62. Drobny, N. L., H. E. Hull and R. F. Testin. "Recovery and Utilization of Municipal Solid Waste," U.S. Environmental Protection Agency, Solid Waste Management, SW-10C (1971), p. 42.
63. Midwest Research Institute. *Study of Processing Equipment for Resource Recovery Systems, Vol. 1*, U.S. Environmental Protection Agency, Kansas City, MO (December 1978) pp. 55–58.
64. Midwest Research Institute. *Study of Processing Equipment for Resource Recovery Systems, Vol. 1*, U.S. Environmental Protection Agency, Kansas City, MO (December 1978).
65. Palumbo, F. J., M. H. Stancyzk and P. M. Sullivan. "Electronic Color Sorting of Glass from Urban Waste," Bureau of Mines Technical Progress Report 45 (October 1971).
66. Allen, H., S. Natof and L. C. Blayden. "Pilot Studies Processing MSW and Recovery of Aluminum Using an Eddy Current Separator," in *Proceedings of the Fifth Mineral Waste Utilization Symposium*, E. Aleshin, Ed. (Chicago: IIT Research Institute, 1976), pp. 161–168.
67. Haynes, B. W., S. L. Law and W. J. Campbell. "Sources of Metals in the Combustible Fraction of Municipal Solid Waste," Bureau of Mines, RI 8319 (1978).
68. Raymus, G. J., and E. H. Steymann. "Handling of Bulk and Packaged Solids," in *Chemical Engineers' Handbook*, 5th ed., R. H. Perry and C. H. Chilton, Eds. (New York: McGraw-Hill Book Co., 1973), pp. 7-1–7-50.

REFERENCES

69. Renard, M. L. "Design Considerations #69 for Municipal Solid Waste Conveyors," in *Municipal Solid Waste: Resource Recovery*, U.S. Environmental Protection Agency, EPA-600/9-81-002C (March 1981), p. 30.
70. Wilson, D. G., Ed. *Handbook of Solid Waste Management* (New York: Van Nostrand Reinhold Company, 1977), p. 126.
71. Lisiecki, H. G. "RDF Storage and Retrieval Problems Cause-Effect-Options," in *Energy Conservation Through Waste Utilization*, Proceedings of 1978 National Waste Processing Conference (New York: American Society of Mechanical Engineers, 1978), pp. 199-206.
72. Vesilind, P. A., and A. E. Rimer. *Unit Operations in Resource Recovery Engineering* (Englewood Cliffs, NJ: Prentice Hall, Inc., 1981).
73. Bastian, R. E., and W. R. Sesman. "The Design and Operation of a Chemical Waste Incinerator for the Eastman Kodak Company," in *Energy Conservation Through Waste Utilization*, Proceedings of 1978 National Waste Processing Conference (New York: American Society of Mechanical Engineers, 1978), pp. 557-568.
74. Huang, W. C., and D. L. Nelson. "Duluth Co-Disposal Facility," in *Proceedings of 1980 National Waste Processing Conference* (New York: American Society of Mechanical Engineers, 1980), pp. 551-556.
75. Grant, R. A., and N. A. Gardner. "Operating Experience on Combined Incineration of Municipal Refuse and Sewage Sludge," in *Conversion of Refuse to Energy*, IEEE Catalog No. 75CH1008-2CRE (November 1975), pp. 280-286.
76. Scaramelli, A. B., et al. "Resource Recovery Research, Development, and Demonstration Plan," The MITRE Corp., Bedford, MA, MTR-79W00173 (1979).
77. Thome-Kozmiensky, K. J., Ed. *Recycling Berlin '79, Vol. 1* (Berlin: Springer-Verlag, 1979).
78. Thome-Kozmiensky, K. J., Ed. *Recycling Berlin '79, Vol. 2* (Berlin: Springer-Verlag, 1979).
79. Browne, F. L. "Theories of the Combustion of Wood and its Control," Forest Products Laboratory, #2136 (December 1958, 1963).
80. Hecht, N. L., D. S. Duvall and B. L. Fox. "Investigation of Advanced Thermal Chemical Concepts for Obtaining Improved MSW-Derived Products," U.S. Environmental Protection Agency, EPA-600/7-78-143 (August 1978).
81. DiNovo, S. T., et al. "Preliminary Environmental Assessment of Biomass Conversion to Synthetic Fuels," U.S. Environmental Protection Agency, EPA-600/7-78-204 (October 1978).
82. Spano, L. L. "Enzymatic Hydrolysis of Cellulosic Wastes to Fermentable Sugars for Alcohol Production," paper presented at the Clean Fuels from Biomass Sewage, Urban Refuse, and Agricultural Wastes Symposium, Orlando, FL, January 27-30, 1976.
83. Apple, J. M. *Plant Layout and Materials Handling*, 3rd ed. (New York: John Wiley & Sons, Inc., 1977).
84. "Materials Recovery System—Engineering Feasibility Study," National Center for Research Recovery, Inc., Washington, DC (1972).
85. Hecht, N. L., et al. "Design for a Resource Recovery Plant," paper prepared for Amax, Inc., University of Dayton Research Institute, Dayton, OH (1974).
86. Sussman, D. B. "Resource Recovery Plant Implementation: Guides for

Municipal Officials Acounting Format," U.S. Environmental Protection Agency, Solid Waste Management, SW-157.6.
87. *Eng. News Record* (June 20, 1974).
88. VanNess, M., Jr., Ed. *Proceedings of the Seventh Mineral Waste Utilization Symposium* (Chicago: IIT Research Institute, 1980).
89. Vesilind, P. A., Ed. "Summary Statement," Engineering Foundation Conference on Municipal Solid Waste as a Resource, Duke Environmental Center, Henniker, NH, July 1979.
90. Committee of Tin Mill Products Producers. "State of the Art—Ferrous Metal Recycling in the United States," American Iron & Steel Institute, Washington, DC.
91. Velzy, C. O. "Energy Recovery from Solid Wastes: Opportunities and Problems," in *Proceedings of the 1980 National Waste Processing Conference* (New York: American Society of Mechanical Engineers, 1980), pp. 1-4.
92. Alvarez, R. J. "Status of Incineration and Generation of Energy From Thermal Processing of MSW," in *Proceedings of the 1980 National Waste Processing Conference* (New York: American Society of Mechanical Engineers, 1980), pp. 5-26.
93. Soldano, L. P. "Ferrous Metals Recovery at Recovery 1, New Orleans, Louisiana," project summary, U.S. Environmental Protection Agency, EPA-600/S2-81-102 (July 1981).
94. Soldano, L. P. "Recovery of Aluminum from Municipal Solid Waste at Recovery 1, New Orleans," project summary, U.S. Environmental Protection Agency, EPA-600/S2-81-121 (July 1981).
95. Archer, T., and J. Huls. "Resource Recovery from Plastic and Glass Wastes," project summary, U.S. Environmental Protection Agency, EPA-600/S2-81-123 (August 1981).
96. "Resource Recovery Activities," *City Currents*, special issue (March 29, 1982).

INDEX

Aerobic digestion: biological treatment of waste, 75-76
Air classification: material separation, 101, 111; zig-zag unit, 132-133
Alcohol: fermentation from cellulose, 10,12; production from waste, 74; ethane production, 77
Amax, Inc.: model plant design, 105-107,128-133,135,137
Ames, Iowa: plant, 7
Anaerobic digestion: biological treatment of waste, 14,76
Aspiration: separation processing for light nonferrous fraction, 111

Baghouse: air pollution control, 101
Bacterial fermentation, 76-77
Bioconversion: cellulose conversion, 10; ethane production, 77
Biological: treatment of waste, 75-76
Black Clawson Company: pilot plant, 7,17
Bureau of Mines: pilot plant, 63,107; product recovery flow plan, 17

Capital cost: model plant design, 140
Carbon char production: produced through chemical treatment, 72,75
Carborundum Co.: Torrax process, 12
Chicago, Illinois: resource recovery plant, 7
Coal replacement: pellets, 8
Codisposal: 33,70
Combustion Equipment Associates (CEA): proprietary process, 10
Compaction of municipal waste, 29-30
Conveyors: pneumatic, bucket elevators, vibrating, and screw, 64-68

Corrosion: waste storage units, 103
Cyclone: deairing during processing, 111; used as a settling chamber, 54-55

Dade County, Florida: plant, 7
Densification: extruders, 69; of ferrous fraction, 111-112
Densified RDF: characteristics, 8-9
Density: physical characteristics of MSW, 29,51; separation, 58-59
Dept. of Energy: project, 14
Depolymerization: thermal decomposition, 71
Dryers: used in processing refuse, 73
Dust: control systems, 43; in waste processing, 101

Eastman Kodak Co.: sludge combustion, 71
Eco-Fuel II: process flow plan, 10-11
El Cajon, California: San Diego County demonstration project, 12
Electrostatic Devices: used for separation, 63
Electrostatic Precipitators: 5
Environmental Protection Agency: format for economic analysis, 137; performance levels, 5; project, 5
Ethanol Production: bioconversion, 77
European waste facilities: codisposal by combustion, 70-71; plant facilities, 2,21
Explosion: suppression equipment, 132

Ferrous separation, 109-110
Flailmill: used as shredders, 42

157

Flakt Company, Sweden: flow chart, 18; process, 16
Flow plan: plant design, 82-83, 90-91, 108
Franklin, Ohio: pilot plant, 7,17; fiber recovery process
Front end: recovery process, 5
Froth floatation: magnetic separation method, 58
Fly ash, 5
Furnace design: 2; moving grate, rotary kiln, 70

Grizzly screens: 44-45
Glucose production: from waste decomposition, 74
Gravity separation (see froth floatation, air classification)

Hammermills: 9, 41-43, 109
Handpicking: refuse separation, 109
Harrisburg, Pennsylvania: incinerator, 71
Heat content: of waste, 33
Hempstead, New York: plant, 7
Hydrolysis: cellulose conversion through fermentation, 10, 12
Hydropulping, 7

Incinerators: continuous feed, traveling-grate, reciprocating-grate, ram-feed, rotary-kiln, 2; moving grate, 9; residue, 4 (see also, modular incinerator, water-wall incinerator)

Japan: waste processing, 2
Johnson City, Tennessee: flow diagram, 20

Lane County, Oregon: waste recovery plant, 7
Liquefaction: of cellulose, 12-13
Los Angeles: methane recovery, 14

Manpower costs: model plant, 144
Magnetic separation: 109-110, 132; pellets, 9
Magnets: belt or drum, 60-61; secondary cleaning and ferrous processing, 109-111

Market requirements: data, 95-96; steam production, 97
Methane: production and recovery, 14
Milwaukee, Wisconsin: plant, 7
Modular incinerator: 97
Montgomery, Ohio: seasonal variations in waste stream, 26-27
Morphology: waste particles, 42; characterization, 30-32

New Orleans: resource recovery plant, 7; trommel data, 49

Occidental Research Corporation: 12 process flow plan, 13
Odor: control, 2, 103
Oil: combined with refuse, 9
Operating costs: model resource recovery plant, 141
Optical sorting: 62-63
Organic fines: recovery, 110
Organic liquids: used in froth floatation separation, 58

Paper recovery process, 16-17
Particulate removal systems, see electrostatic precipitators, 5
Pellet fuel: production, 96
Pompano Beach, Florida: plant flow plan, 16; project, 14
Powdered refuse derived fuels: 9-10, 33
Precipitator: performance, 10; temperature, 5
Processing lines: single or multiple, 99-100, 113, 124, 140
Putrifaction: storage problems, 103
Pyrolysis processes: 12; bioconversion of cellulose waste, 10

Resource recovery plants: closures, 7; plants in operation or under design, 150
Refuse derived fuels: 5-8, 10
Residue disposal: 4

Saccharification: 76-77
St. Louis, Missouri: resource recovery project, 5-6; data, 107
Screening: fines removal from fuel fraction, 110

Seasonal effects on waste stream: 24–27
Shredding: 5–6
Sink-float process: gravity separation, 58
Sludge incinerators: 70
Sorting refuse: 96; manually, 109; optically, 62–63
Steam generation: 2
Stoker unit: furnace design, 4
Storage bin design: 67–68

TVA: trommel study, 49
Torrax process: carborundum Company, 12
Trichoderma Viridae: bacterial fermentation, 76–77
Triple/S Dynamic Company: design for processing, 49
Trommels: first-stage waste processing, 7, 129, 132; screens, 44–50; angle of incline study, 49–50

Union Electric: RDF project, 5
University of California: trommel study, 49
University of Dayton: 135; plant design, 105–107

Vibration: waste separation by density, 58–59

Water-wall incineration: 2–4, 96–97
Weigh station: incoming trucks, 124–125
Wet classification: rising current separation, 112
Wet Process: paper making, 7
Wet Pulper: size reduction of waste, 43
Windrows: open field composting, 76
Worcester Polytechnic Institute: 13, 15; flow plan, 75
Wright-Patterson Air Force Base, Ohio: D-RDF demonstration program, 8